Ellen Taaffe Zwilich   Mark O'Connor   Jane Ira
Bloom   Tan Dun   Joseph Schwantner   Samuel
Adler   Richard Stoltzman   Evelyn Glennie
Wesley Jefferson   Kenny Loggins   Jim Messina
Paula Robison   Bob Dorough   Sarah Chang
Paavo Järvi   Mischa Maisky   Pamela Frank
Erik Nielsen   Mike Gordon   Midori   William

DISCARD

Parker   Gregory Turay   Randy Weston
Susanne Mentzer   Sharon Robinson   Jaime
Laredo   Marcus Roberts   Mickey Hart   Tilla
Henkins   Sharon Isbin   Christopher

Yousuke Miura   Eugene Skeef   Libby Larsen

Water Music

Water Music

PHOTOGRAPHED AND ORCHESTRATED BY

MARJORIE RYERSON

Sixty-six renowned musicians from around the world celebrate water in words and music

INTRODUCTION BY PAUL WINTER

University of Michigan Press ANN ARBOR

Published in the United States of America by

The University of Michigan Press

Printed and bound in Spain by Bookprint, S.L., Barcelona

♾ Printed on acid-free paper

2006 2005 2004 2003   4 3 2 1

A CIP catalog record for this book is available from the British Library.

LIBRARY OF CONGRESS CATALOGING-IN-PUBLICATION DATA

Water music : sixty-six renowned musicians from around the world celebrate water in words and music / photographed and orchestrated by Marjorie Ryerson ; introduction by Paul Winter.

   p.   cm.

ISBN 0–472–11338–0 (cloth : alk. paper)

1. Photography of water. 2. Ryerson, Marjorie. I. Ryerson, Marjorie.

TR670.W38 2003

779'.3—dc21   2003044771

For more information on the Water Music project please visit the web site at www.WaterMusicProject.com

Designed and typeset by Mike Burton in Monotype Garamond and Trajan fonts.

Color separations by PreTech Color, LLC., Wilder, Vermont.

The net royalty income from the sale of this book is being donated to the United Nations Foundation, to be used to support water as a resource for the Earth and for its inhabitants, now and for future generations. The United Nations Foundation has created the "*Water Music* Fund of the United Nations Foundation" for all donated revenue from book sales, as well as for donations to the fund from other sources. In the awarding of all of its grants, the foundation encourages the United Nations to work in partnership with non-governmental organizations. The foundation expects that a variety of non-governmental organizations, both international and local, will work in partnership with the United Nations in the implementation of these water programs.

FOR

Bob Beyers

A DEEPLY ETHICAL JOURNALIST AND FRIEND

WHOSE UNCONDITIONAL SUPPORT, GENEROSITY,

INTELLIGENCE AND HUMOR HAVE

INDELIBLY ENRICHED

MY LIFE.

PREFACE

*T*he making of *Water Music* began in the mid-1990s, when I increasingly found my camera lens turning toward water. Initially, I photographed water that I happened upon in the course of my travels. But I quickly grew dissatisfied with leaving these encounters to chance, and began to change the itinerary of my trips to include and then to exclusively photograph water.

I was enthralled by the challenge of capturing on film the astonishing breadth of ways in which water presents itself—what I came to think of as water's *essence*. Yet the more I photographed water, the more I came to realize that water's appearance knows no limits.

Water has many faces, many forms. Sailors crossing Lake Superior just after sunset can watch the lake's surface turn from deep blue to gold, copper, blazing red, dusty pink, pewter and—finally—black. Water mirrors archetypal forms: perfect spirals form on the inside edges of waves; tiny precise vortices trail from the end of a canoe paddle's stroke. Water in flood brings a kind of uncontrolled rage to everything it touches; water in quiet lakes and streams lures and nurtures all living beings.

Water can even be surprising. It is a relatively clear substance that often draws its colors from the surrounding environment. Yet water can also glitter with colors absent in the landscape. On a damp and dark, foggy day in Maine, the waters I captured through my lens were swirled with streaks of bright turquoise; on a blue-skied and cloudless, sunny day, the deep waters of Washington State's Puget Sound looked exactly like polished black marble. Pebbles tossed onto a slow-moving tributary of the Mississippi River create rings in the water that expand outward in perfect circles, in spite of the river's current. And in spite of the fact that the river is flowing downstream, the growing rings seem to remain anchored in place. As they expand even farther, the many rings pass over and through one another, yet somehow retain their shape. It is almost as if the rings and the river are separate entities.

Even now, thousands of negatives and many years later, I continue to encounter visual manifestations of water that are both mysterious and like nothing I've previously seen. Watching water can be like witnessing magic.

When I first decided to put some of my water photos into a book, I knew that I wanted to create a book that could give back to water for all the years of pleasure and fascination it had provided me, both professionally and personally. I wanted to create a book that could assist in the protection and restoration of water in many different parts of the world. Now, through a partnership with the United Nations Foundation, the book's charitable recipient, *Water Music* will be able to do just that.

For this book, I wanted an exceptional text. Who could better speak for and about the most essential and lyrical of elements, I thought, than musicians, who themselves so lyrically and essentially enrich all of our lives? So I set out to invite some of the greatest musicians of our time, asking each of them for their intimate impressions of and responses to water.

The many musicians on these pages have made space in the midst of their demanding lives to contribute to the book, generously donating thoughts, poems, essays and music about water for the book's text.

It is my hope that the collaborative joy behind the making of this book will be communicated on its pages, and that the respect for water inherent in this project will ripple outward into the world.

Marjorie Ryerson

THANKS

$\mathcal{I}$ am grateful for the many kinds of support I received during this book project from friends, acquaintances and total strangers who assisted in many ways to help me complete this large undertaking. A few of the key people behind the success of this book are listed below. I wish to extend thanks as well to the many other people whose names are not listed here but who also lent a hand in a variety of large and small ways throughout the many years it took to bring this book into print.

Chris Hebert *and*
Phil Pochoda *of the*
University of Michigan Press
Michael Burton
Joe Citro
Mason Singer
Richard Saudek
Laura Blake Peterson
Kelly Going
Sylvia Moss
Gay Gaston
Sharon Robinson
Bill Stetson
Barbara Ligeti
Emily R. Jones
Nicholas Jones
Revell Allen
George Konnoff
Chinweizu

Julia Starzyk *and*
Synneve Carlino *of the Chicago*
Symphony Orchestra
Aimee Petrin *of the Flynn Theatre*
Mike Peluse *of the Vermont*
Symphony Orchestra
Cem Kurosman *of Blue Note Records*
Dennis Bathory-Kitsz
Patricia Morris *and the staff at the*
Dartmouth Music Library
Krista Ragan
Nancy Tabor
Mary Elizabeth
David Garten
Robert Gershon
Jonathan English
Lydia Petty
Shane Heroux
William Ryerson

Yoko Nakashima
Bob Paquin
U.S. Senator Patrick Leahy
Peter Fox Smith
Sandra Duling
Tom Edwards
William Ramage
William Carter
Gail Ellison
John Ruskey
Brian Vachon
Marie Kittel
Andy Jaffe
John Fago
Lar Duggin
Bobbi Perez
Rob Gordon
Bryan Billado
Jonathan Czar

Alexia Borenstein
Brian Calderara
Jennifer Usle
Jay K. Hoffman
The Trustees *of the*
Vermont State Colleges
Bonnie Tangelos
Sadahei Kusumoto
Kazuto Ohira
Eriko Fukuoka
Dan Denerstein
Onyekachi Wambu
Cenovia Cummins

Mariko Hancock
Jeremy Lesniak
Keith Wheeler
Jack Byrne
Frederick Johnson Pianos
Bruce McDonald
Tim Fallon
David Rahr
Vince Crockenberg
Susan Delattre
Beth Monturoi
Hannah Jeffery
Steven Rockefeller

Thanks to
Castleton State College
in Castleton, Vermont,
for the many ways it has supported me
throughout this long and complex
project.

————————————————

And unequivocal thanks go to the
many musicians on these pages and to
their agents, managers and friends
whose support and dedication made
the creation of this book possible.

This book exists with the generous support of

THE LAMSON-HOWELL FOUNDATION

CASTLETON STATE COLLEGE

THE HOROWITZ FOUNDATION

VERMONT PURE SPRINGS

MATHEW RUBIN OF EAST HAVEN WINDFARM

CONNIE DeWITT AND ISABEL & HAROLD DeWITT

&

THE UNIVERSITY OF MICHIGAN PRESS

And with the indispensable assistance of

THE CHICAGO SYMPHONY ORCHESTRA

THE FLYNN THEATRE FOR THE PERFORMING ARTS

CONTENTS

INTRODUCTION

By Paul Winter

"The sacred water" is the name given to Siberia's Lake Baikal by the Buryat inhabitants of its shores. Its ineffable beauty has drawn me there seven times. Baikal is the world's oldest and deepest lake; it contains 22 percent of the world's fresh water, of unequalled clarity and purity, as well as many extraordinary endemic species, such as the world's only freshwater seal. Accordingly, Baikal has been the subject of many scientific studies that have attempted to measure, probe and plumb its majesty, but nevertheless, its mystery and beauty remain unquantifiable. Russian people as a whole revere the lake as a sacred site, to such an extent that when industrial pollution threatened Baikal, they rallied against the Soviet government in its defense, and Russia's environmental movement was born. Such is the powerful allure of water. It represents an aspect of wild nature without which we die.

Earth is the only planet in the solar system with abundant liquid water. Newly created, Earth was a hot, airless ball of rock, but as its crust cooled, outgassing from its interior produced water, which stayed in liquid form and pooled into the seas. As the Earth continued to cool, the atmosphere condensed and torrential rains added to the oceans, which now cover two-thirds of the planet's surface and contain 97 percent of all the water on Earth. The ice caps of Greenland and Antarctica hold another 2 percent,

and only 1 percent, without which most life forms could not exist, is in the ground or in the air. Subject to the forces of this most powerful earth shaper, the land continues to be molded, spread and sculpted by rushing rivers, swelling seas and tremendous strokes of rain. Our planet's total water supply is believed to remain constant, recycled as it is evaporated from the oceans, condensed, precipitated, and drained back again to the seas — the same supply that existed three billion years ago.

Three hundred and seventy million years ago our vertebrate ancestors came ashore, yet still our young begin their lives in little salty seas of amniotic fluid. How profoundly I experienced that connection to our watery origins when my daughter emerged from her mother's body into the warm water of a birthing tub and into my hands. She was born at home, but increasingly hospitals and birthing centers are making water-birth available to laboring women, recognizing that, beyond its pain- and gravity-relieving effects, water can play an important beneficial role in releasing inhibitions and encouraging surrender to the birth process.

Though we know our lives depend upon water, ironically we take for granted this most precious resource, rarely pausing to consider the ways in which we squander it. Fifty thousand gallons of water, for example, are needed to make rayon for a living-room carpet; forty thousand gallons are needed to produce the steel for one car. The market for home water-treatment systems and bottled water is now burgeoning, as scarcity of clean, accessible sources has rendered water even more "exploitable."

As we pollute our water sources, we poison all life systems. In the past, we revered the magical properties of water. This strange substance, self-transforming and transformative, was understood to contain as well as to enable life. Recently we have been able to appreciate that life-sustaining power in a new light: the Bovis scale scientifically measures the vitality of water and can compare the depleted energy of treated water, forced under high pressure through straight pipes and out of faucets, with the vitality of wild water, whose numinous life energy Marjorie Ryerson so exquisitely portrays in *Water Music*. This extraordinary work commands our awe and respect; her unique perspective opens our eyes again to water's miraculous beauty and life-giving power. *Water Music* is a gift for our times.

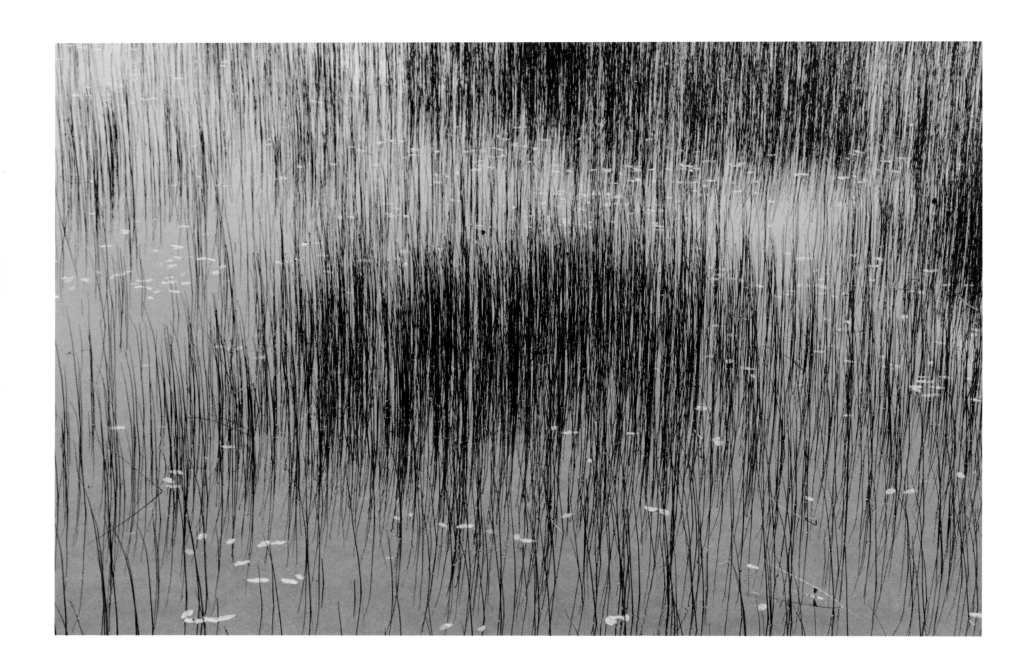

CAROL MAILLARD

H_2O Flow

A tear /

drop of rain

tapping rhythm from an old faucet

leaks from the corner of my right eye

frozen patterns crystallized and infinitely

complex / sparkling on a winter windowpane . . .

it

washes the sand from my sun-toasted toes

and

seduces me to jump in for one more lap . . .

it is sweetly fresh

flowing from an upturned jug

to parched lips / calming / nourishing /

refreshing and

filling me with

pure spirit . . .

cool lakes

tender brooks / are

music to my inner ear / and

I am lulled into a luscious dreamland

where

I dance to the waves . . .

H_2O-flow

is music

vibrating in my cells

is music

on a rain-kissed tin roof in St. Maarten

is music

of crying clouds and roaring waterfalls and fierce ocean blues

it

moves into quiet crevices

expands / contracts / expands

moving rocks and trees and then

gently rests

wherever . . .   however . . .

Purification,

ritual baths, healing forces

flow thru me

blessing and increasing

seducing / concealing / cleansing / healing /pacifying /releasing / rocking and warming and

serenading

ALL THAT I AM . . .

Watersong . . .

DAVID HARRINGTON

Water is a primordial substance that perpetually renews itself. Moving continually among its different natures, water is essential to the flowering of all life on earth. On the surface of each drop of water one can imagine seeing reflected the mirror image of every form life has yet taken. Water has seen it all, its molecules have been through it all, and yet water silently continues to hold the secrets of its immense past. Perhaps the next raindrop once moistened the eye of a browsing dinosaur. Wouldn't it be amazing to be able to know the entire history of a single drop of water, a bead of sweat or a tear?

NADJA SALERNO-SONNENBERG

There is a certain rhythm to water.

The pulse of the wake beneath you and the timing of the ripples caused by

a pebble or insect.

If you see it, then feel it.

You can hear music that has not yet been written.

BRUCE COCKBURN

In the mid-seventies, I was living on a well-watered patch of land an hour or so southwest of Ottawa. One spring, '75, I think, the runoff lasted a long time, and seemed to me particularly musical. Standing on the deck of the house at night, you could hear water singing on every side. Somehow that seeped through my brain and came out in the form of the guitar piece for which this is a sketch. The piece was recorded and appears on an album called *In the Falling Dark*.

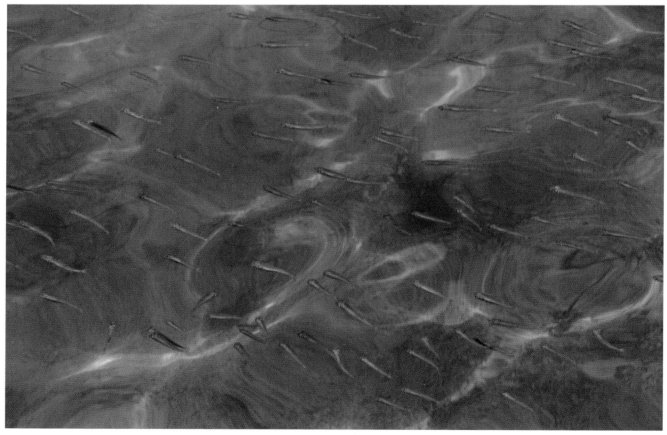

BOBBY M^CFERRIN

in the deep end
all my fears show.
i don't swim, i
flail. i let go
of faith in God
and fear is knot-
ted in my bod.
my lungs are scared
of liquid air.
i don't swim i
flail out there.

my ocean is sound,
the audience, waves.
i swim the deep
when i'm on stage.
i show up, show off
wing it and sing
and fling my fears
to faith which brings
to deep water
as Jesus said
that's where the catch
is. but i dread
to swim in there
but faith would dare.

give me danger
in B♭ major
swimming around
that joyful sound
my feet firmly
on the ground
flailing, wailing
sinking, sincere
while drowning
in music
faith, water
and fear.

MARY ELIZABETH

I

In the darkest part of the night, too tired and too scared to sleep, I went to the river. The drive was quick and effortless at that hour—no traffic to interrupt me as I glided down Vermont Route 117. I parked in a gravelly pull off, pulled on my hip boots, and trudged along the edge of the cornfield, thinking of Robert Frost's lines, "And when I come to the garden ground, / The whir of sober birds / Up from the tangle of withered weeds / Is sadder than any words." But is it sadder than the shadowy weeds in the night silence before the birds awake, I wondered?

I stood on the bank, alone, poised. A momentous moment. Nobody in the whole world knew where I was. Standing on the bank, my feet on rocks, my body surrounded by air, I could not feel myself as a force in the world. Utter isolation. But what a difference to stand in the air and to stand in the water! The water noticed me as soon as I stepped in. It parted around me; it welcomed me with sound as it lapped against me. As soon as I entered the river, I could see how my presence mattered, how what I chose to do changed the flow of things. When I entered the water, I joined the fish, the raccoons, the ducks, the heron, the moose, the rocks, the leaves, the twigs—and the water accepted all comers. When I enter the river, I become part of the unity of water in all the world because there is only one water.

I played with the river. I went downstream to a tiny falls and added myself to increase and change the effects. I took handfuls of pebbles and tossed them in the air to land in the water, trying different heights, different timing, different

amounts. Was this noise, sound, music? How could one tell? Was it good for the river to have more pebbles? Was it bad?

Trying to exist in the moment—trying to put aside analysis, pattern-making, prediction—I watched the sky and the water as the sun rose, the sky lightening, first to white and only then bluing, as color returned to the world out of the darkness and shadow into which night casts everything. The sun dappled and lapped the water, the water shimmered and shadowed and rose and broke against my legs, and with light and color came more sound as the birds awoke and the fish began their morning dance, and the animals came to quench their morning thirst.

At evening, I went back to the river, again to the Winooski, but to a different spot by a bridge. I sat on a rock near the waterline, recently visited by a raccoon who'd left her crayfish shells. I let my bare feet hang into the water and watched the color become deep, intense, and passionate—both above me in the sky and below me on the water, till I was surrounded by sunset blaze—and then slowly drain from the world, leaving it barren and black. Except . . . except for one warm, yellow light from a farmhouse across the river, lighting the night and giving it a glow that said "hope." When it was dark, I went home to sleep for one or two hours, knowing that when I woke up, I would go back to the river to begin again.

II

Early spring flooding tossed up huge chunks of ice, some thirty feet from the riverbed where I had watched the sun rise. I would have to climb over them to reach the river, and I hadn't anticipated such difficulties in just getting to the water, and my feet were cold in my hip boots. Then, suddenly, one chunk shuddered beneath my feet, and shattered, not into fragments, but into long, sturdy, beautiful crystals. I had known the minute beauty of the snowflake, but never something born of water this big and bold and diamond-like. And the feel of the collapse, too, was worth repeating—not just a fall into space, but an almost orderly caving in as the joints between the crystals disconnected and each rolled loose from its neighbors. I stayed with the frozen water that day, savoring the breaking open of the ice into diamonds—I never went down to the river. And I have never experienced that phenomenon again.

III

My neighbor, staying the night in his camp a few feet from the Lincoln River, heard a knock on the door at 2 A.M. "Get out of your house. The river's coming." He got out. Barely. The flooding river unleashed the power of water and swept away a chunk of his house, including his slate-bottomed billiard table (never seen again), and destroyed the value of his property. I went to help clean up. We found some oil paints (the river took the billiard table, but not the oil paints?), a bicycle, and other things that the river had decided in its wisdom not to take, and I found a portion of a hipbone of some creature (a cow?) that the river had decided to leave behind. I have the hipbone in my room beside a piece of driftwood, both reminding me of water's power and of unfathomable choice.

IV

I have written between four hundred and five hundred songs, and I've discovered that writing a song is fundamentally an act of unity. This being so, I cannot commit words to music if I cannot in some way, on some level, hold them as my own, because the joints between the words and the music should, I believe, dissolve like the boundaries between one drop of water and another, until there are not lyrics over here and music over there, but only the song. This does not mean that they stand in one particular relationship, the music illustrating the words, or the music supporting the words, or any such thing—simply that they have become an organic whole: they ebb and flow as one. And I do not feel *constrained* by the parameters set by the words; rather I am free to be myself in their context. But I must be careful, too, because I hold someone else's work in my hands, and it is precious. So it is a unity with respect, a unity that does not diminish either, but enhances both: freedom and unity—the Vermont state motto. There is unity in the river, but it has no freedom: it takes whatever we give and joins with it. So for this reason—beyond its beauty, beyond its life-giving—the river, too, we must treat as precious.

RANDY NEWMAN

I have loved rivers since I can remember. I don't know why.

I lived in New Orleans as an infant and perhaps had some spiritual encounter

with the big river they have there. Perhaps it's spending most of my life

in southern California, which is, as you may know, quite dry—a desert in fact

had we not stolen water from elsewhere. Everything but the chaparral is fake.

The Willamette, the Russian, both Colorados—it excites me just to see them.

I find them (I wouldn't include either Colorado here) the most beautiful things in the

world.

The feeling I first had as an eight year old going to a night ballgame in Los Angeles,

going through a tunnel and coming up on a green field lit up like an emerald—

that's what I feel when approaching and then seeing a river. Cities that have them

(no matter how they may have been desecrated) are blessed.

You can fish in Portland and you could and should swim in the Merced.

You set your clock by the tide. You asked it when the wind would change, you woke or slept by its coming and going. You gave it your days when the boats were moved, your nights when the fish came down the bay.

GORDON BOK

You'd awake on a day whose name you forgot, but you'd always know the state of the tide and how far it would come or go. You measured your sleep and your work and your ways by the ways of the tide; it dogged your watches and kept you content. And there is no adding the hours and days, the years spent waiting for the tide to serve.

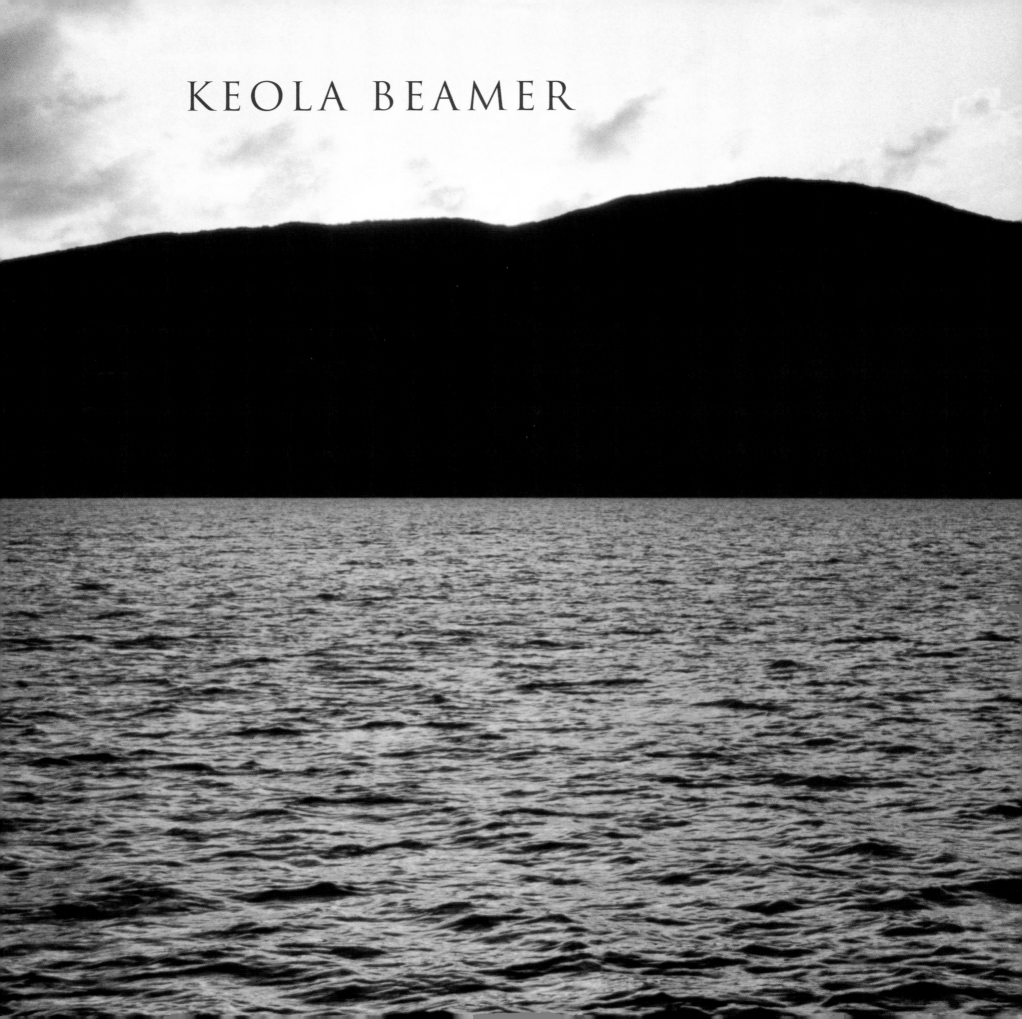

KEOLA BEAMER

I live in the little town of Lahaina, Maui, with my wife, Moanalani. Recently, we returned from a trip to witness the Obon festival at the Jodo Mission. Lahaina is a multicultural little town. We have learned to respect and appreciate each other's beliefs.

Our Buddhist friends prepared a hundred paper lanterns with the names of their deceased relatives. In each lantern, a candle was lit, and after the sun went down, the lanterns were set out into the warm, flowing waters of the Pacific.

As we watched beneath the stars, the old priest chanted on the beach and the little paper lanterns rode the waves. The lanterns rose and fell, carried by the soft winds and currents of the Pailolo Channel. Each lantern sparkled in the vast darkness of the ocean, journeyed forth and then finally expired, the twinkling lights extinguishing in the distance, one by one.

To me this was a metaphor for our own lives. We each shine briefly in our journey, and then one day, we too expire. Standing on the cool sand and watching the lanterns being carried by the sea, I heard a bittersweet music in my heart. It was the music of life and loss, a reminder that we are only here on this earth for a short time. We need to embrace our family and our friends, to laugh and love and make the most of each day.

HERMETO PASCOAL

Water

is the most beautiful and positive

shadow of the Earth.

TAJ MAHAL

Sky Juice

I awoke in the back seat of my parents' '46 Chevy. Are we home yet? No, said my mother, not yet. Soon! I rubbed my eyes. It was dark now, and when I had fallen asleep the sun was still going down. Now a big silver moon was playing hide 'n' seek with us in the sky. And then it happened! We passed a large lake, and the silver moon displayed her beauty across the face of the rippling lake. That first experience sent a little chill through this then five-year-old's body that made me hold my breath. I called it "Big Shimmering Glimmering." And I have called it that ever since.

Although I get to travel so much and have seen this natural phenomenon all over the world, it still never fails to give me the same wonderful, awesome feeling every time I behold its beauty.

Any body of water—pond, lake, stream, river, bay or ocean—starts my mind wandering. I cannot help but be drawn to it and to think we Earth people have so much to be thankful for. I know we cannot take this for granted.

Countless songs, and some of my most popular songs, have water at the heart of their theme: "Fishin' Blues," "Light Rain, Light Rain," "Rain from the Skies," "The Sky Is Cryin'," "Backwater Blues," "High Water Everywhere" and so on.

Where does water come from, I asked my parents? This time Dad said, From the sky, and we Caribbean folks call it Sky Juice.

NERISSA NIELDS

Snow

There is nothing as forgiving as the first snow of the year. It arrives via an eerily wet-looking sky and blankets the half-raked yard in late November or early December, covers the untidy skeletons of last summer's perennials, leaving the landscape pure and clean and honest, at least on the surface. It falls silent as a prayer, gentle as the caress of bow to violin strings, consistent as the beat of a drum, the beat of a heart. It is a silent symphony in the morning, the birds temporarily muted by the stillness, blinded by the white.

Best of all is the promise that there will be some reprieve for the child in each of us, some hope that whatever unfinished business we might have, whatever English composition, geography test, math quiz that might have been scheduled for the day, has been wiped off the calendar while we, innocent again, spend the morning letting the flakes fall onto our faces and tongues, riding our toboggans down Breakneck Hill, making angels in the sweet, white powder, crunching snowballs between our mittened hands. And those of us old enough to remember the heat of the summer, the thirsty, dusty fields, say a silent thank you and close our eyes in relief, knowing the snow is white gold which will slowly seep into the ground, joining its own deep in the earth to fill the rivers and streams in March.

And we who make music, we who write, bow our heads as well, holding our hands together to make a cup, asking the snow to fill us up with its frozen baptism, drenching us with a dry watering so that what lies deep and still and seedlike within us can begin to stir to life and make its way, slowly, up to the surface.

ALAN GAMPEL

The Sounds of Water:

Roaring waves,

Gurgling springs,

Sparkling dew,

Cascading falls,

Droplets of rain,

Exploding fountains,

Rippling streams.

Resources for my keyboard palette,

Music to my ears.

IAIN MAC HARG

Taigh nan Uisgean Binne
House of the Singing Waters

Jig

Iain Mac Harg

2nd Time 4th Part

1st Time 4th Part

MARY YOUNGBLOOD

Liquid Poetry

My relationship with water began like everyone else's, with a flood of amniotic fluid that both proceeded and followed me from the womb of my mother to the womb of *the* mother, Mother Earth, a wondrous place where music would eventually find this Aleut/Seminole girl, joyfully claiming her spirit, heart and soul.

Since then, water's influence and inspiration have been poignant and powerful in many significant areas of my life, from baptism to the hiss of fresh water being poured from a wooden ladle over hot rocks in a sweat lodge.

I was born under the sun sign of Cancer, its element, of course, being water. As Cancer is one of the more mysterious zodiacal signs, the emotions of those born under it are known to ebb and flow from great heights of mountainous bliss to the deep darkness of dread. In short, I was a song just waiting to happen.

Adopted at seven months old and raised in Kirkland, a suburb of Seattle, I learned to swim in the cool green waters of Lake Washington. As a young girl I took to the water like a fish, my parents will tell you.

Coming full circle in my adulthood I learned to kayak in the even colder waters of Puget Sound, naturally taking to the sport like those who had walked before me. The Aleuts are well known for their beautiful, sleek kayaks and skill with the traditional skin boats.

To gather inspiration for my second album, *Heart of the World,* my percussionist Steve took me into the Sierra Nevada foothills near my home in Northern California. With flutes, drums and rattles on our backs, we hiked far up into a gorge of the Yuba River, ancestral land of the Maidu People. Then, in a cool and sacred place, sitting in the soft sand beneath the shadows of massive boulders, I wrote "Yuba" and "Ladybug Dance." It came as a surprise to me that ladybugs "swarmed"! Literally thousands of the little red and black creatures carefully flew by us. The sound of our instruments bounced off the canyon walls, harmonizing with the soft sounds of river water as it hungrily rushed forward in search of the sea, joining our celebration and discovery of a new song.

In July of 2002, I was given the opportunity to perform in Juneau, Alaska, via a weeklong "powwow" cruise. Floating slowly through the glassy smooth channels of my birth mother's people, I realized I had come full circle. I had arrived at the home of the Chugach Aleut, my people. It was a treasured pilgrimage to a place I had never before known, a strangely familiar place, a place called home.

I lazily watched the reflections of these ancestral channels swirl by, in constant flux and movement, and was reminded of the parallels of my own life, and of my journey as a musician, native woman and mother: gentle waves, weaving their way in and out of my destiny, stirring the heart with rich texture and contrast from shore to shore, seemingly coming from nowhere, yet going somewhere—always in search of a song.

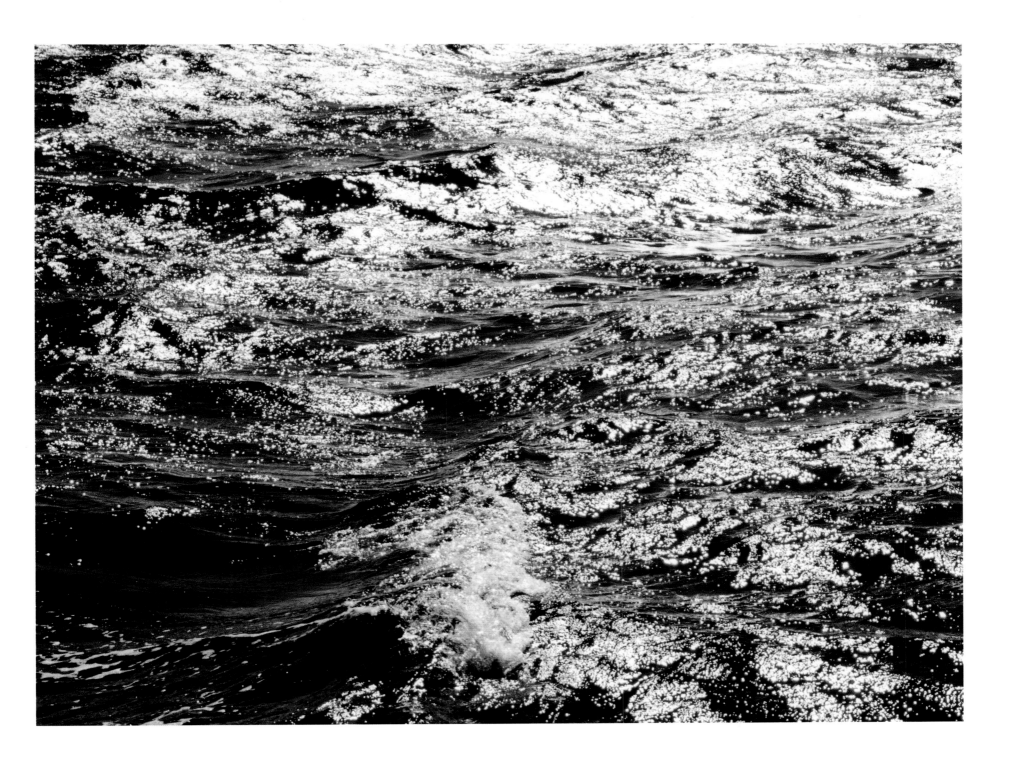

TUNDE JEGEDE

The Waterpot

Thirst brought me
From the dusty field,
Women in the gardens
Men gone to town,
Only chickens
Quietly pecking earth.
In a dark corner
Stood the earthen waterpot,
Its surface still
As glass

I dipped the cup
In the deep well water
And cast a shadow
Of darkness.
Into my reflection
The sad, old whisper
Of a man distorting my face

I am Water
Source of life
I nurture the fields
And bring green wisdom,
But Death comes
To reap my harvest

I am Water
Source of life
That serves
The Ancient Baobab,
But Death comes
To uproot knowledge

I am Water
Source of life
That feeds
The toiling people,
But Death comes
To steal the song

I dropped the cup
And fled, with unquenched thirst

RENÉE FLEMING

Water is

> The splashing brook of florid, majestic Mozart

> Cheerfully running melismas

> The flow of a river, a long Straussian phrase seeming never to end

> The majesty of the ocean—a chorus of untold voices which makes us seem small
> and our sorrows trivial

> The calming repetition of waves—a soft mournful drone which slows time and breath

Water and great music enable us to step out of our daily lives for a moment and experience timelessness. We are humbled by the experience and grateful for the reminder that life is fleeting and the soul wants feeding. Mostly when I think of water, I think of flow—the same flow I feel in my body as a tone begins and travels into aural liquid on breath, perhaps meandering or chromatic, an ever-flexible waterfall of sound.

RUSSELL SHERMAN

Before the Big Bang there was, we are told, a cauldron of boiling foam, prelude to the birth of gods and man. Out of that trembling sea the stars were born. Nothing can come forth unless triggered by such convulsions, spasms of the soul, waves of light or feeling. This fluid mix, viscous enchantment or demon, brings forth both poems and universes. Shelley suggests that the poem written down has already lost some of the feverish urgency that spawned it. And we, humans, have arisen from the oceans; such was the nature of our prenatal home, a maternal ocean born of tears and trembling, adapting us to the fluid rhythms of life.

Each composer preempts an indelible portion of the four elements. Crudely speaking, Mozart discourses in fire and air, Brahms is more earthy, while Ravel and Debussy are more overtly the protagonists of water. But these stereotypes demand a price, for the very language of music is expressed in waves, swirls, swells and eddies, and even the transparent articulations in Mozart are like brilliant whitecaps which geometrically arrange the sea. We respond then with emotion, whether trickles or waves of emotion.

Rachmaninoff said that the pedals are the soul of the piano. The damper pedal creates layers of sound where long and short sonorities may mingle. The dark, opaque undertow is matched by the glittering foam of the surface. Rippling sounds cavort on a grid of harmonic pillars that form a kind of "submerged cathedral." And the pedal, ideally as graceful as the maneuvers of a dolphin, discriminates among these layers of deep and light. Without it the piano would be a fish out of water.

Water, paradoxical and addictive, womb of human life and source of Narcissus' fatal obsession. Water, which reconfigures light and color. Water, architect of miraculous designs and of the endless melody so seductive to Wagner. Water, which can never be stepped into twice, and therefore becomes the paradigm for performances that are singular and unrecoverable.

It is water that eternally and helplessly rearranges its components, and that reminds us of the special bond between music and life. For they both are fleeting, ephemeral and magical. And, finally, it is the intrigues and intricacies of water that precede and illuminate our destiny.

BRAD MEHLDAU

Flowing water and music

both teach us something about time.

They remind us that time isn't a series

of enclosed moments, but something more constant.

"Beginning" and "end" are us—

our own fictions, our own mortality.

PATRICIA BARBER

Let It Rain

lord, let it rain
i can't stand the lie of a blue sky
one more day
can't you make that pitter, patter
sweet teardrop splatter
against my windowpane
c'mon, bring down the sky
let those clouds and me have a good cry
let it rain

lord, let it rain
take away the glare of the empty
shape of the day
can't you make that liquid fly, bubbles rise,
heatwave die,
that water slip and slide me away
your sun makes me feel like a fool
god your weather just don't match my mood
let it rain

let the thunder crack,
let the lightning snap the tension of a
long, dry afternoon
wash away the time,
cover me in shadow, then forgive me for
lying in bed,
listening to the
rain fall on my roof

lord, let it rain
i can't stand the sound of the silent sun on my face
can't you make those downtown hopping,
grocery shopping,
perky, plodding, cheerful folks
go away
c'mon bring on the flood
let my soul have its day in the mud
let it rain

c'mon bring down the sky
let those clouds and me have a good cry,
let it rain
lord, let it rain

VLADIMIR ASHKENAZY

I'll never forget my first impression of the sea

    —as a child I remember the feeling of eternity

       when I saw the endless expanse of water.

I don't really know with what I could compare that first impact—

perhaps only with my first experience of hearing the symphony orchestra:

in the sense of a strange belonging to the elements—however diverse

they might be in this particular instance.

GARRICK OHLSSON

I wish I had gills.

Ever since I was a baby (according to family lore), I could somehow swim easily and with pleasure. Even if lore exaggerates, being in the water, in addition to being on or around it, has always felt natural to me. As a kid, I was a champion underwater swimmer. I regretted having to come up for breath; I even insisted on keeping my eyes open in chlorinated or salt water. Underwater it felt quiet, free, vividly three-dimensional. (I know this sounds like scene 1 of *Rheingold*. Maybe I was a Rhein Junge in a previous life.)

I was reminded of these childhood feelings a few years ago when I took a rafting trip through the Grand Canyon on the Colorado River. For much of the trip the water was too cold for swimming, but being on it in such surroundings made the trip one of my all-time great vacations. We slept on the riverbank and made many side excursions. My favorite occurred at Havasu Falls, a large inlet that feeds the Colorado. Walking in about a half-mile we found a series of small falls with lots of opportunities for wading and swimming in deliciously tepid water.

The most blissful moment occurred when I swam underwater into a cave just big enough for two or three people. Inside was a comfortable, air-filled natural dome. Light from outside illuminated it softly, revealing mineral colors of green, blue, and turquoise. Plant roots sought moisture by reaching down from the roof, taking root beneath the water in the sides of the cave. In this fossil-rich part of the canyon, I was astonished to see that these mineral-encrusted roots were fossilizing as I watched. They still retained some flexibility, but their calcification was clear to the touch. Every kid loves to have a secret place—an attic, a tree house. This grotto, full of watery magic, would be my place. If I had gills I would live there.

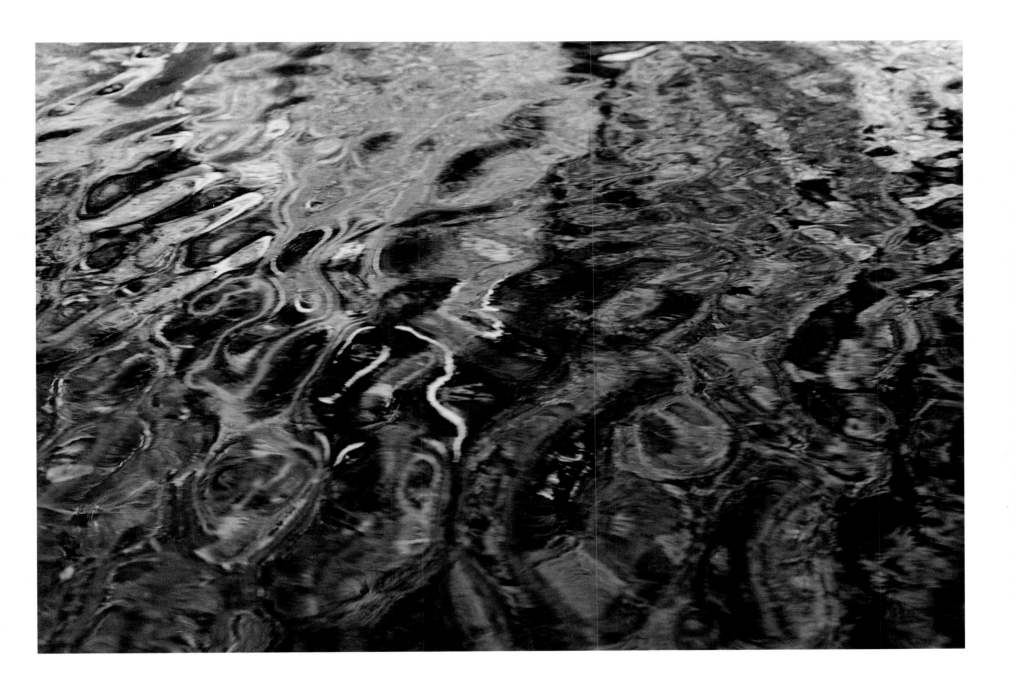

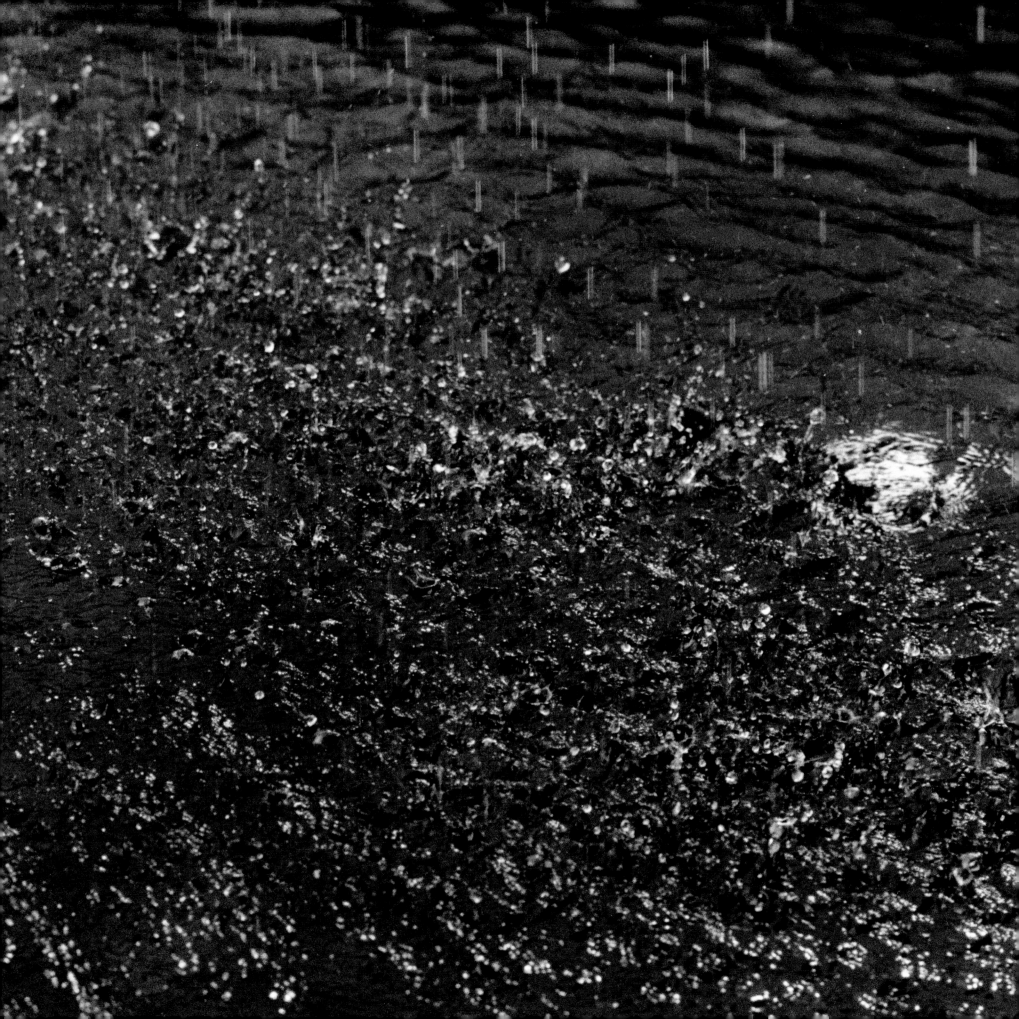

BASIA DANILOW & PETER SANDERS

Starting with the smallest trickle of rain it begins the journey. Joining forces with the nearest young brook or stream it gains strength. Along the way, each rock and branch that it crosses tells a special tale as it ripples and gurgles. It is growing.

Next it winds its way into the nearest river. Always majestic in its seasonal beauty, the river can be calming and peaceful—or perilous, as rapids form around the bend. The travels continue.

Just as its many cousins do, the river finally opens into the waiting sea—a vast expanse embracing the river's many stories. The journey is almost finished. In time, evaporation pulls the droplets upwards toward their final destination. The cycle begins once more.

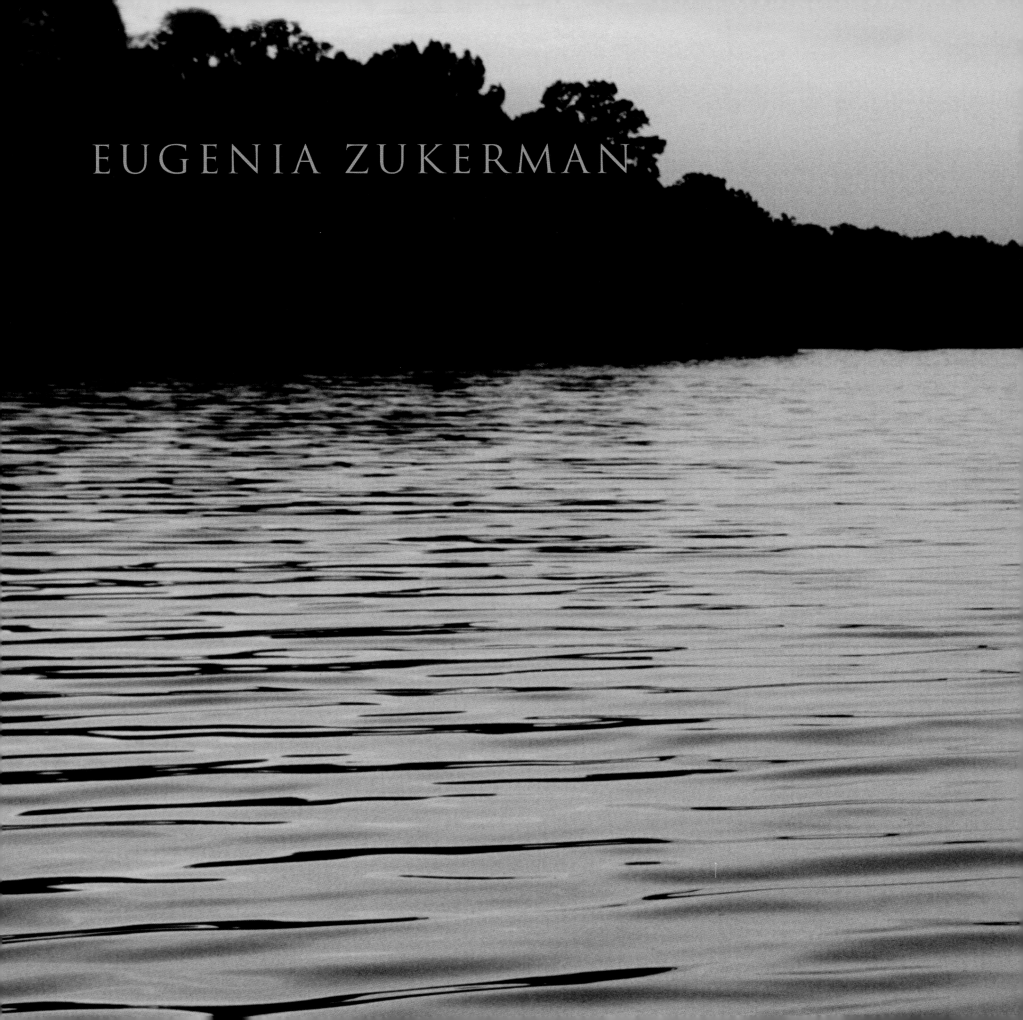
EUGENIA ZUKERMAN

Splashing in a pond on a hot summer's day, floating and feeling buoyant and free—what happier memories can we have? Water is as necessary to us as it is desired by us. We are lulled by it, nurtured by it, protected by it, drawn to it, and fascinated by it.

I have lived next to a great body of water for nearly thirty years. The mighty Hudson River flows past my windows, and not a day goes by when I do not thrill to the morning light on the water, or a little red tugboat pushing a big barge, or a sailboat flitting by like a white moth, or a sunset turning the sky to rose then purple then lapis lazuli.

The music of water is endlessly intriguing, whether it is the arpeggio of a waterfall, the improvisations of a stream, the crescendo of a wave, or the roulades of a brook.

The sound of an oar cutting through the water, or a trout leaping up into air, or a dragonfly flitting on the surface—these are the instruments of a liquid symphony that resounds in our souls. No wonder composers over the centuries have tried to capture and distill those sounds, with their *barcarolles* and *ondines* and *reflets dans l'eau*.

Even the terrors of water add to its allure. Neither my near-drowning as a child, nor the currents that caught me and nearly dragged me out to sea as an adult, can keep me away from water. I will always seek it out, need it, want it.

To float weightless in water is to return to a prenatal world without worry or care. To gaze at water is to see infinite variety and possibility. Water flows beneath my windows, and it brings me solace, strength, and inspiration. It is a conduit to my hopes and dreams, a daily reminder of the power and beauty of life.

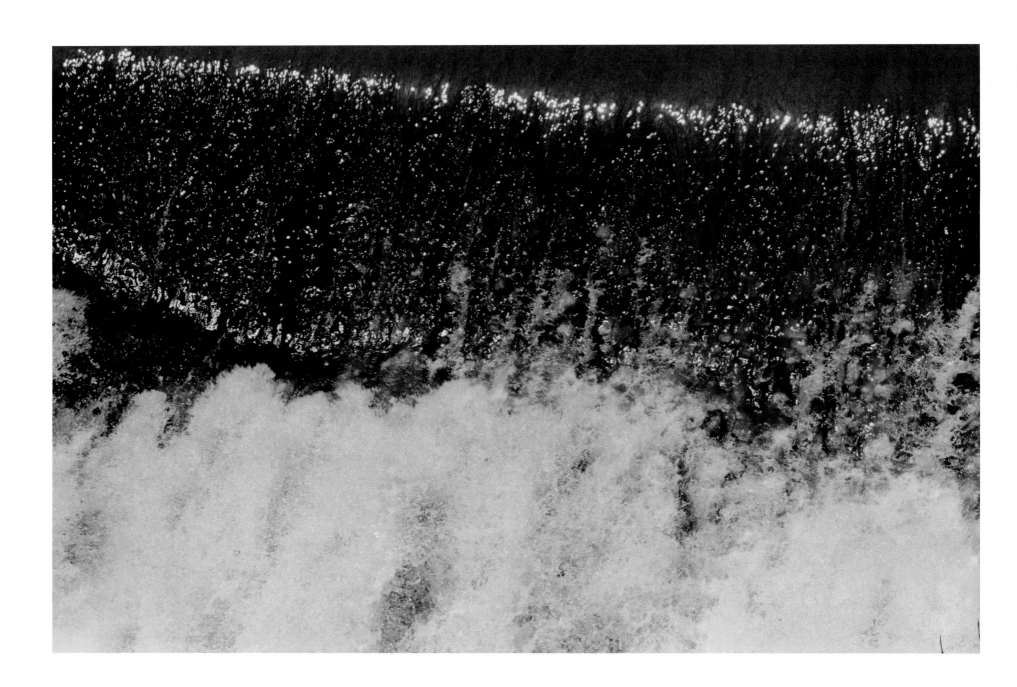

PETE SEEGER

KEB' MO'

Power and Humility

Water, essential to our survival.

Strong, worthy & nurturing. It can save your life or kill you.

It can carve out valleys or wear away mountains.

But water, even with all its power and magnificence, always

goes to the lowest place it can find. It waits to be lifted up

by the rays of the sun and to be transported by the wind.

Then it rains down upon us.

This cycle of transformation is like life & death. Death being

just another part of life, whereby our spirit is lifted and our

bodies transformed.

I myself am mostly made of water. And we are all made of

light.

I am one with all. Therefore the water is me and I am the water.

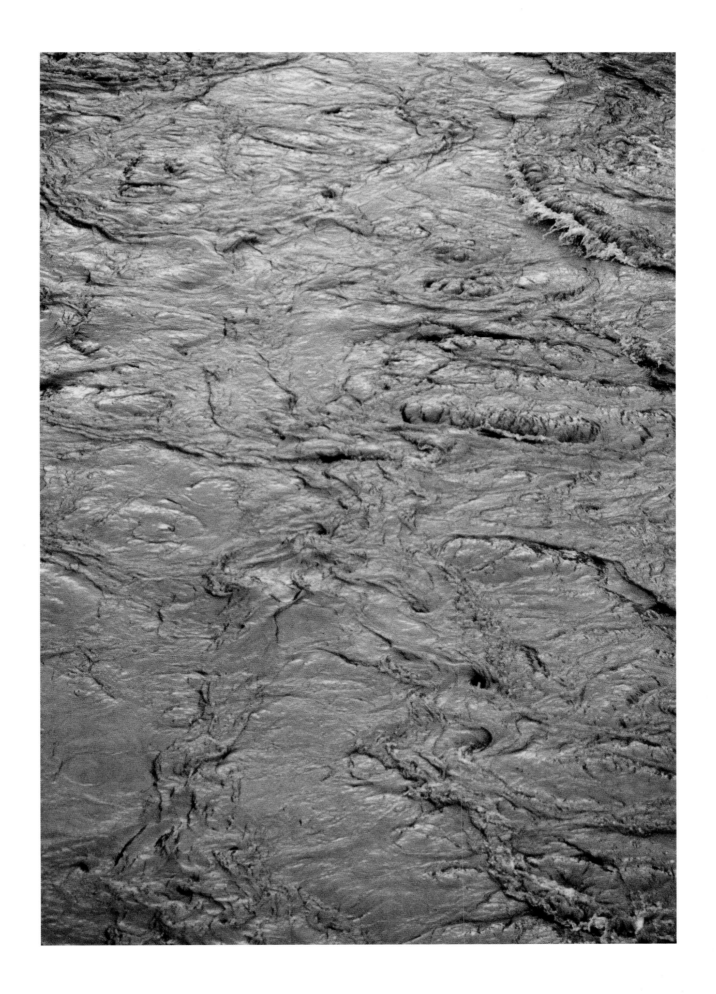

JOHN RUSKEY

Build a bed and lay your head and be seduced in the blood murmuring through the mud, the pulse & heartbeat of the earth, the charge & discharge of thunderstorms and the chaotic mixing of moisture & air, ground down and retuned by the tireless rhythm of the river on its return journey to the sea—the wheels of the gaiic cycle turning the wheels of the soul, the medium being returned to its resting place while the gaseous other is lifted out of the oceans and made to replenish the high places in freshets & showers & torrents—and then again is lowered to its resting place—brought together in arroyos & rivulets, coulees & bayous—the wheel an eye—the motion an eddy—the power transformed—energy made to change directions—the horizontal made vertical—spokes the lines of power directing motion to and from the madly spinning whirlpool, the gears of the titanic transformer, finely tuned and forever flowing—man setting his wheels in the current and siphoning a portion of the power to grind his grain—the bear swatting fish out of rapids—deer placing their lips to drink and making concentric ripples emanate outward in dusky pools—the current becoming one with the other greater — the glass refilled—the battery charged—the ions and special shape of two hydrogen and one oxygen triangulated—the terrestrial trio—forever flowing, freezing, and being evaporated and flung again into the wild whimsies of the wind and the actions of the atmosphere—the dead made to revolve & spin, to beckon cryptically—unseen hands making the inanimate alive—the vertical motion of the trees made horizontal—approaching winter solstice by blooming boils of liquid death sliding endlessly from the north—arctic air suddenly spilling into the great valley of the mississippi and making the waters of middle america congeal into a mathematician's mosaic—solid parallelograms—vanishing points—commensurate crystalline stars along the muddy banks—and those incommensurate—the imprint of

heaven's geometer—seeing the great extension of lines and figures flung out impossibly extended over the arc of the sky, too extended to comprehend in a single glimpse—and yet each point, each star, each blade of ice a part of the whole, recognizable as one—the toe of the ballerina cantilevered impossibly beyond the balance of the flung head and the reach of the hand—each point creating the geometry of the whole—seen spinning from the tip of the toe to the bridge of the nose—the axis of ice—at turns vertical—horizontal—then a star—vertical spine—a star—a line—and then an explosion of lines and soft supple motions carried by the bones—the tree felled in the forest—the web-spinning spider—the woodpecker tearing rotten chunks from cottonwood trunks—the blades of ice each being carefully placed in fractal patterns of crystalline structure, and then destroyed in the spring—solid again become liquid—the waters of the world never at rest—some circular motion always making the corner, cutting the bend, carving the oxbow, completing the circle, curling, contemplating the curve, coming to a spiral, whirling & swirling, dervishing, boiling & eddying, whirlpooling, erupting, devouring, lapping & curling & frothing, haystacking, revolving, helixing & bubbling, pushing & pulling, sucking & exploding, outbursting—the motions that mirror the impulses of the universe and make accessible the forms with which cosmic cartographers and artists are enabled to decipher and then to scribe the secret movements—the secret movements which are heaven's design and simultaneously the germ of all organic life on earth—a 2,300-mile-long waterfall the mississippi born from a marshy pond deep within the great north woods—a tidal channel the length of the continent, the serpentine trunk of life and productivity—the stem upon which and by which the same creative impulses which concocted the universe are observed, expressed, recapitulated, and brought to conclusion in foamy climaxes—and

69

then reassembled, themes again repeated—hardened over time—protecting the flowing that has been started and allowed to continue—sheltering the sweet haven of love and productivity, the rosy cambium—the warm core of our tree—from which sprang she and he—and from which we extend ourselves and to which we are always connected—the arpeggios and crescendos of a cold watery orchestra carrying on the tradition of universe construction—the one who laid the geometry of the stars surely is the same hand who scored the water—the patterns of the horse's head and the orion nebulae are retraced momentarily and then dashed into the chaos of madly sliding motion—flung apart by its own motions & inertia—later reclaimed by the muddy moving mass to be de-threaded in swirling motions, and then organically recombined, the pieces seamlessly juxtaposed into writhing helixes—where the sparkling reflection of the setting sun ends there the gloom begins—the forests invaginated, coiled-up, and catapulted over undulating forms and mixtures of incandescent pastels and earth tones—the forest de-horizonized—the image made a mirror image of and lassoed over the darkest reflections in the somber masterpiece—fantastically silhouetted in exploding streams of sunshine & humidity—lingering in the undulations, the accents of the most serious wave-motions—the elegant brush strokes of god—blues & purples flowing out of yellow, green, aquamarine, the orange glow of the sun, and streaks of yellow breaking through the holes in the clotted clouds, the tattered shreds of thunderstorms open and close and gyrate around unseen centers of focus, hidden attractors, the yaw of great floating ships in the sky, the yawning of heaven—clouds ripped right out of a bank of fog, or from a cracked cucumber, the criss-cross scales of the sturgeon—skittering along the tops of the trees—the indifferent scrutiny of the glib eye of god

observing mercilessly the passage of pilgrims over her waters and randomly creating chaos in maelstroms in the most magnificent mediums available in the universe—the colors too rich and too full to find in the painter's paint box—one palette spilling over with the liquid tones of water—another with the pastels and gray washes of humidity—the third glowing with the spectrum of the sun—the last piled up with the tints of mud, the sand, the forests—through which all other mediums flow and are affected by—and themselves made to look one way and then later made to look another—the whole combination constantly being tossed against muddy banks like dice—god seen dragging his stick through the mud and the waters following, dancing, sparkling with the sun in pure joy—singing gleefully—madly roaring through the woods and purring into deep pools, the purling whispers emitted from gentle boils, the slapping of snags on the slippery mirror surface—skillfully slicing muddy reality into sublime sluices, snakes slightly submerged and looped over snags—the whistling of willow leaves in the wind—turtles sleeping in the sun—birds & insects making music and the waters heard them and were pleased, and lay smooth in the channel, the boils spreading fat & shiny, the current a gossamer belt of light & colors of the sky, and the sun bathed all in golden warmth, green leaves made yellow, pieces of molten yellow rippling in the waters, a canyon of foliage—the tawny sun melting through shafts of mist, columns of auburn moisture, butter frying in the morning sun and becoming atmospheric—later there were waterfalls of orange, and the forests disappeared into an ochre haze—a thunderhead became illuminated overhead, outlines of gold and flaring red—great moody fields of smooth purple and cinnamon vermilion created in the cloud shadows, ultramarine spots low on the horizon and ashen cobalt in the void—

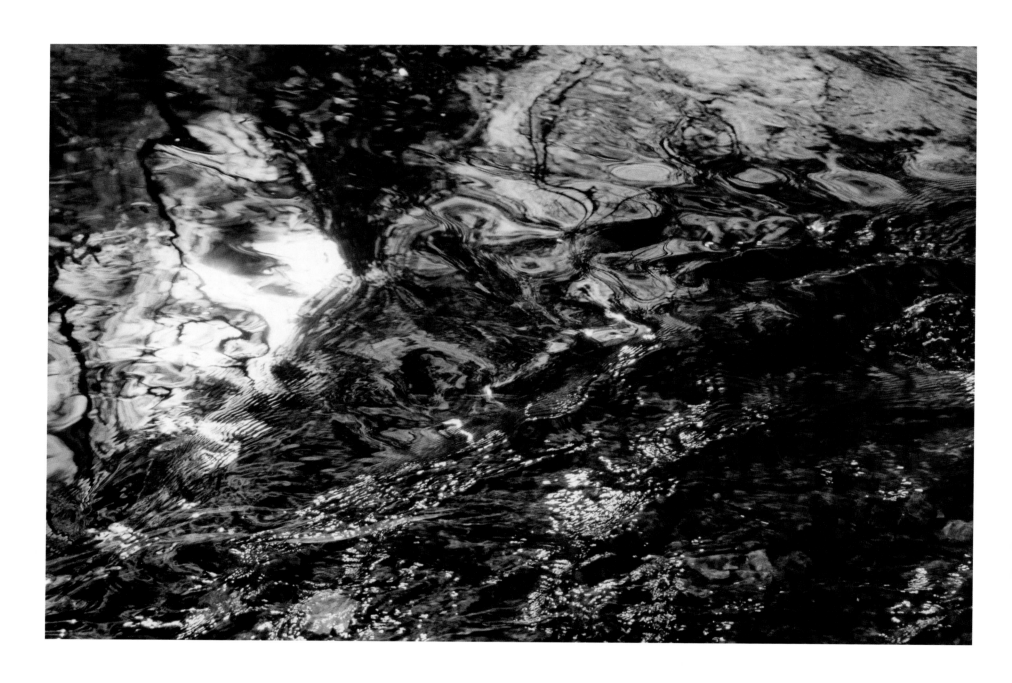

GEORGE WINSTON

I.

Flowing rivers always remind me to keep things moving. The speed and force aren't as important as to just keep moving ahead.

II.

Life to me is trying to constantly keep in balance—it is like balancing on a log in an ever-changing river.

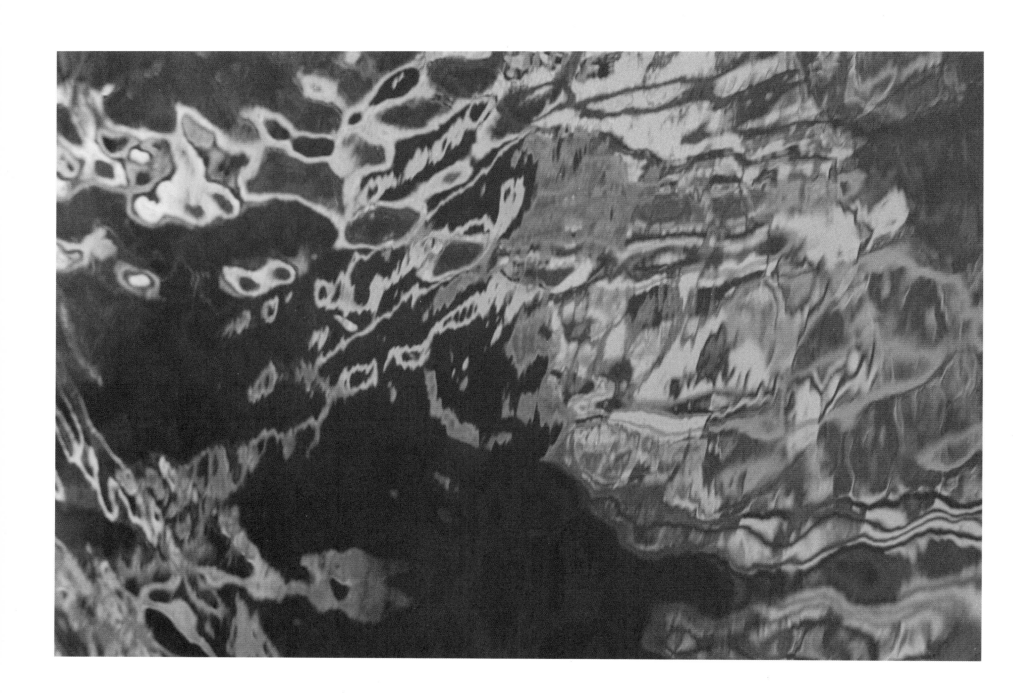

EMANUEL AX

"You never step into the same river twice"—this is especially true for us musicians. We spend so much of our lives in the process of deepening our understanding and, in the process, our love for great masterpieces. The notes remain the same, but the emotions they arouse, and the way we translate the notes, change ever so subtly over time; every performance is a process of change. Perhaps that is why so many composers have been inspired by water—the eternal which is never the same.

DAVE BRUBECK

My house sits beside a stream. It is quiet in the frozen months of winter. Today I am looking at wild and frothing water that roars outside my window, darting around three gigantic boulders deposited there some time in the Ice Age. Hitting a granite ledge, the stream makes an abrupt right turn and rushes away from the house, gushing through a series of little waterfalls, finally spilling into a placid pond. Where the moving water enters the pond, I have built a small studio on an island, and I go there to compose music. And there with the fish and the turtles and the otters, the ducks and the geese as companions, I set to music the speech of Chief Seattle, in which he asks us to protect the sparkling water and the pure air and the trees and plants and all the fish and beasts "for we are but a strand in the web of life." I called that piece, "Earth Is Our Mother." I have since written other music on this little island, and the harmony with nature that I feel at this spot, surrounded by life-giving water, is inspirational to me.

ELLEN TAAFFE ZWILICH

Water—oceans, lakes, my Hudson River—holds a central place in my life and work. For many years now my working space in New York has been dominated by a table pressed against a window with a magnificent view of the Hudson River. The Hudson River, a fiord, actually, is almost like the sea in its many moods. Some days it takes on an almost tropical hue; other days it appears cold and slate-like. The nautical traffic on the river, artery of human activity, is pleasing to see, as are the beautiful palisades on the river's far side. And the light! From my table I can see a reflection of the sunset and, on a clear day, the Tappan Zee Bridge.

In my home in Florida the Atlantic Ocean occupies center stage. In fact, the apartment feels almost like an ocean liner with the sea visible from every room. I love to watch the sea birds and the drama of weather as it plays out on the ocean. Sometimes I need to just sit and look at the sea. I can't say exactly why it moves me or how it inspires my work, but it is clear to me that water has always had a profound effect on my life.

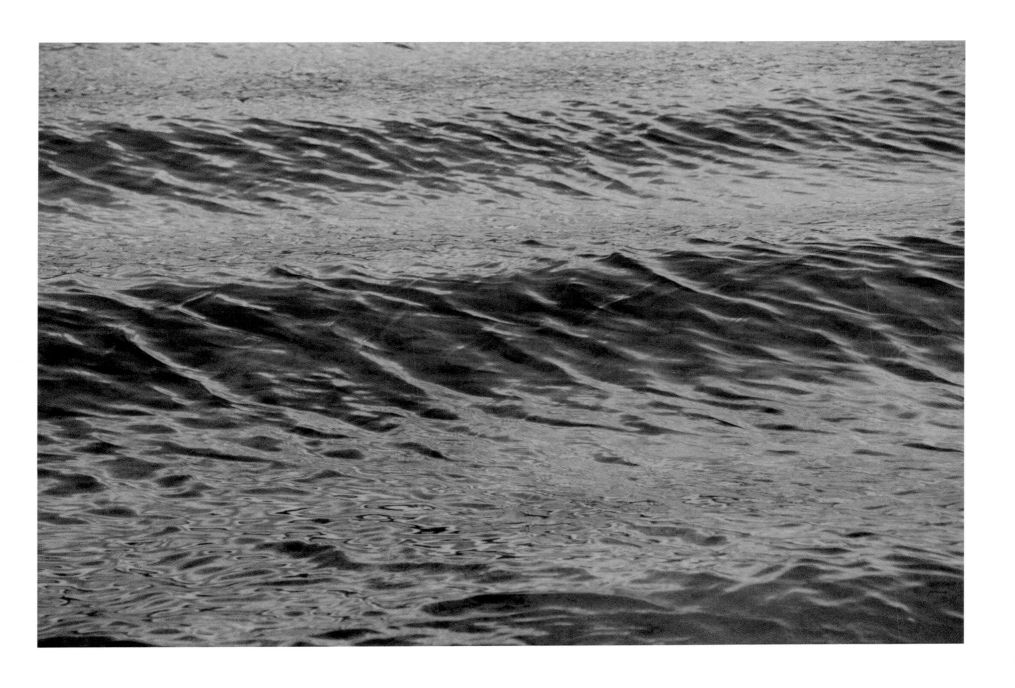

MARK O'CONNOR

Water Music

As I grew up in Seattle, I was surrounded by a multitude of water effects. To the west was that great Pacific and to the east were the Cascades and their rivers, lakes and streams. And from the heavens came the rain. The idle rains.

When I was a kid, this idle rain seemed to me a kind of cruelty. I lived for the outdoors, skating on the streets. Upon the many asphalt embankments of the city, I would ride my board. It seemed to me a perfect daily pendulum swing from the self-isolation of my musical practice room. When I was depressed, when rain would not allow me my childlike release, my speeding ecstasy . . . I felt it was like God taking a constant leak on me, sort of. I don't know why I should have been so surprised by the Northwest, because when I was even younger, my mother told of how I would shoot my toy pistol at the sun, thinking it was not supposed to be there!

Later, when I was on my own, I lived in Tennessee. The place was landlocked, really, and I was unaware of the impact it would have on me. Unlike most West Coasters, I really did enjoy that relentless Tennessee humidity. Whatever moisture there was in the air made me feel good and complete. I found myself building a home on a cliff that resembled a Pacific get-away. I realized that when the Tennessee rains came, it felt like a nagging comfort to me. This was the period when I composed "Idle Rain" for violin and guitar. It dawned on me that I had unknowingly looked for a piece of land to build a home on—that looked good when wet. The water from my childhood was always with me as a backdrop to whatever I did.

Much of my music reflects this water legacy. My favorite tune from my earliest learning on the violin was one my teacher composed, "Midnight on the Water." Even those childhood tears I used to shed because of the idle, irrepressible rain in time would go toward the making of what I now recognize as a bittersweet soulfulness in my music.

These days, things are dryer for me, but I am back living by the Pacific in California. You know, there was really only one good reason why I came back. The water was my connection to serenity, to the best part of my art-life dance I seem to tap out. Maybe an earthbound juxtaposition for my musical creativity.

I compose so many cannons now. Sounds that rush in at the listener in succession, over each other at changing positions, again and again, like waves. Or notes streaming down like water drops whose appearance may seem just alike but, when the falling or the spilling is over, combine to create an accumulation, insights develop and other secrets reveal themselves in the process. In my concertos, my trios, even my newest choral, much of my music is like this motion.

I always say that I love it when I feel the "water in the air." The salt water in the air makes me feel at home now and that feels very fine.

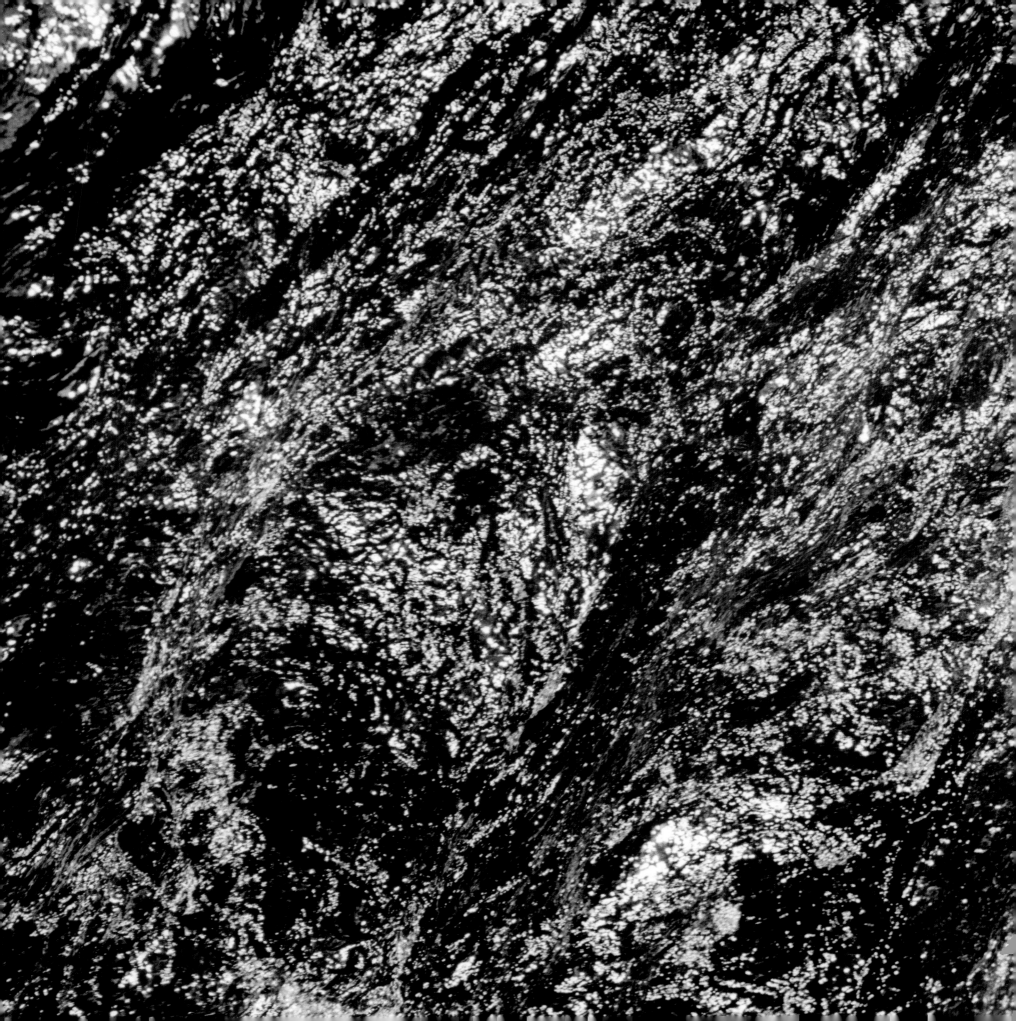

JANE IRA BLOOM

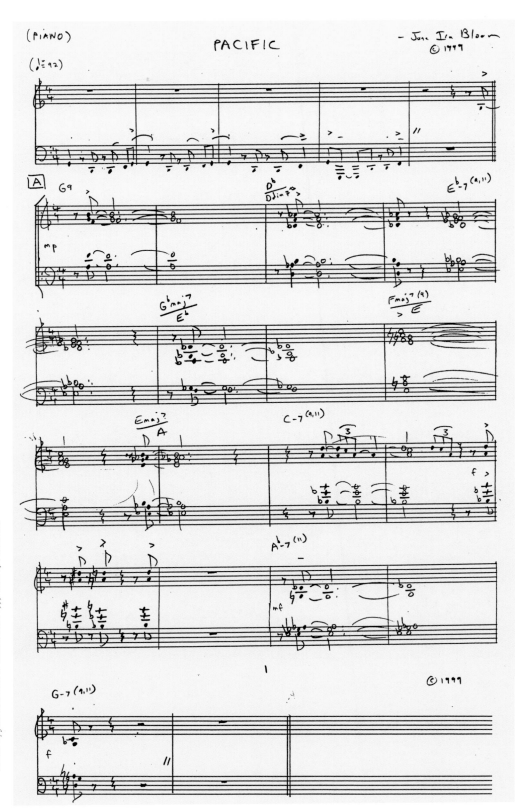

The ocean has always been something mysterious to me. Waves, magnificent distances, and the possibility of exotic adventures always captured my imagination about the Pacific.

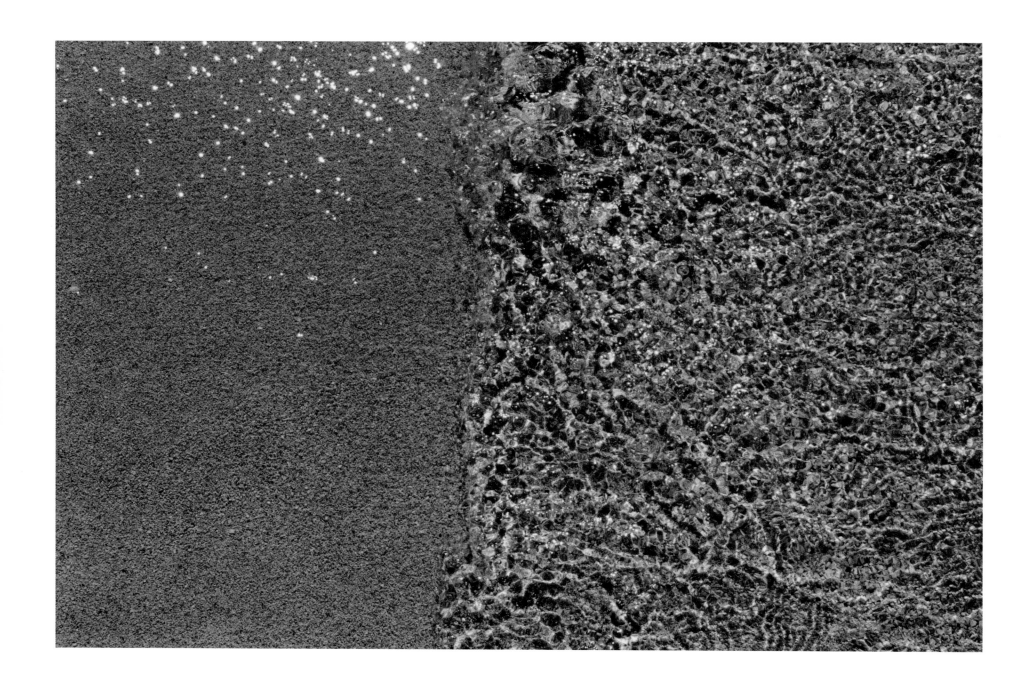

TAN DUN

For most cultures, water is an essential metaphor. It is the metaphor for the unity of the eternal and the external. It plays a major role in birth, creation, and resurrection and for that reason is significantly incorporated into baptism ceremonies.

The life cycle of water replicates resurrection when it comes down to Earth as rain, moves forward through its many living stages on the planet, and then returns to the atmosphere as vapor, where it starts the same process all over again. Resurrection is not only a return to life but also a metaphor for hope, for the birth of a new world, for a better life.

I have composed many pieces of music inspired by water, including *Water Passion after St. Matthew* and *Concerto for Water Percussion and Orchestra—in Memory of Toru Takemitsu.*

The concerto explores the sounds and inspirations of water as well as the use of water itself as a percussive instrument. Some of the many water instruments in the music include hemispherical transparent water basins, water bottles, water tubes with foam paddles, water shakers, water drums (floating wooden salad bowls), water gongs, water agogo bells, and a water phone. In addition to the percussive sounds created by instruments agitating water, the water instruments can play chromatic melodies.

The use of a "Water-Instruments-Orchestra" is musical metaphysics. The expanded color and sound of the orchestra are both inspired by water and inspired spiritually.

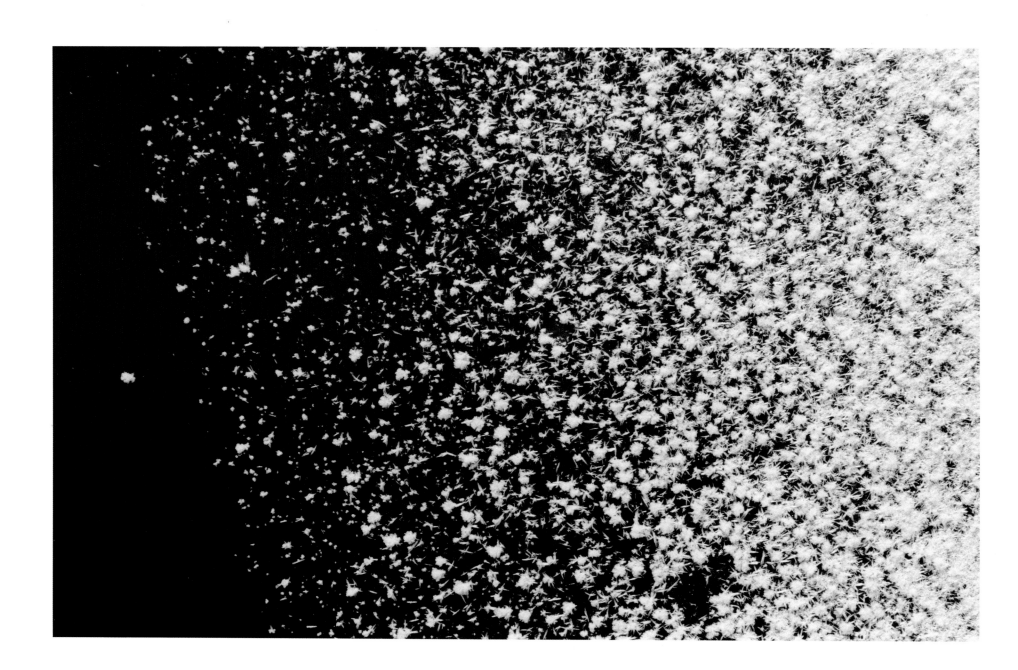

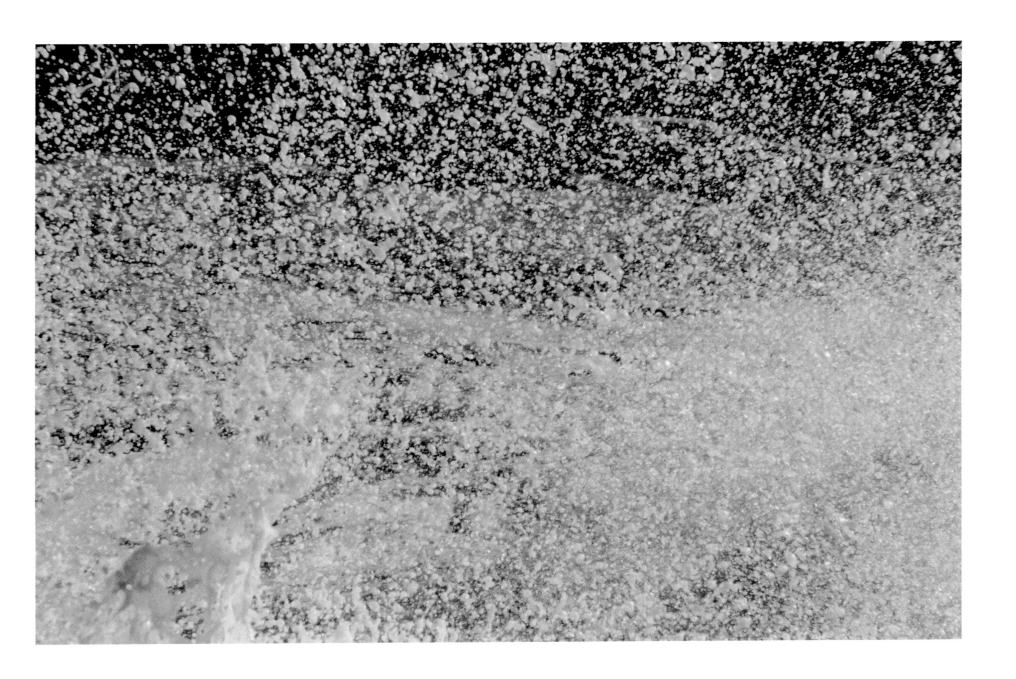

JOSEPH SCHWANTNER

While jogging early one summer morning after a brief rain shower, I came upon an extraordinarily vivid and fully formed rainbow that seemed to appear suddenly and dramatically in the distance. Most strikingly, the rainbow contained sharply etched bands of colors wrapped seamlessly from horizon to horizon. The scale and clarity of shape, the distinctive bands of color and the invigorating clear, fresh, moist air proved to be a potent stimulus to my imagination.

Immediately, I began to consider the possibilities of capturing this powerful experience of water and light, and of framing those elements into a coherent musical design, not only using the rainbow's pure colors but also emphasizing the sense of the refraction, reflection and dispersion of the sun's rays in the mist and falling rain. My composition, *A Sudden Rainbow,* emerged from that dramatic moment. Throughout the composition of the work, the memory of that early-morning experience remained intense and helped to propel the flow and shape of my musical ideas. *A Sudden Rainbow* became the second of four works I wrote for the Saint Louis Symphony while I was composer in residence and living in Chesterfield, Missouri.

The opening of *A Sudden Rainbow,* presented initially by an orchestral choir of winds and percussion, contains an eight-note collection that functions as the primary pitch universe for the music. In addition to winds, brass and strings, the sounds of amplified piano, amplified celesta, harp and percussion collectively form one of the principal coloristic strata emphasized throughout the score, producing a profusion of rich and sharply etched articulations. The music unfolds in a series of stratified orchestral layers that project the material of the work as it spreads outward in bands of musical color, just as the spectrum of light had done in those early-morning raindrops.

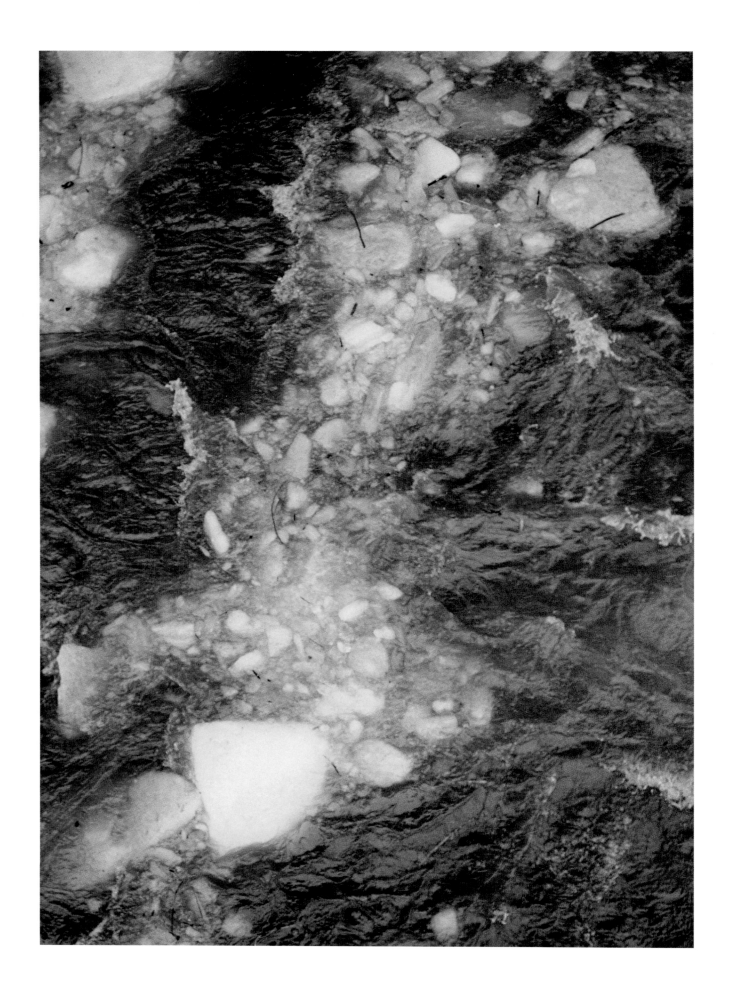

SAMUEL ADLER

When I moved to Rochester, New York, in 1966 to assume a professor-
ship at the Eastman School of Music, I emigrated from the Southwest,
from Dallas, Texas, to this city full of water. Lake Ontario and the Erie
Canal were just two blocks from our house, the Finger Lakes a few short
miles away. How different and almost shocking this water environment
felt.

Two years later, the Rochester Philharmonic Orchestra commissioned me to write a work for its series of concerts called "Places in
America." I decided to put my feelings about my new hometown into
the composition. Since Rochester is a city with enough water to nourish
both its people and its trade, I decided to draw from three folk songs
that each speak of a city by a lake, "Fifteen Years on the Erie Canal,"
"Red Iron Ore," and "The Siege of Plattsburgh." Each song's reference
to water is what appealed to me, and I called the work *City by the Lake*.

The work opens with the feelings generated by the mist of a typical
morning so often experienced in Rochester, that wet, damp, but often
cool and fresh feeling that is so wonderful in the early dawn. This is followed by a fast section that was inspired by my passing the canal every
morning, often in the rain and snow. All that water was invigorating, and
therefore the rest of the piece is full of energy. I wanted, musically, to
point out something that was so vital in my new home, which is that so
much water adds a great deal of wakédness to life, and makes one more
alive than when one lives in the placid temperatures and waterless places
of the Southwest.

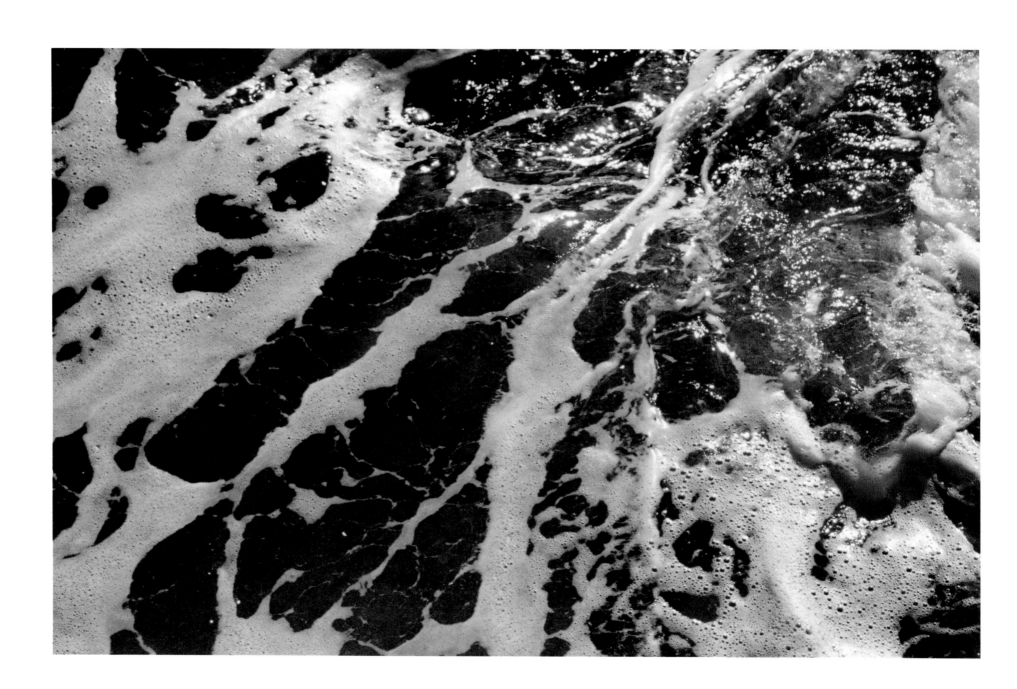

RICHARD STOLTZMAN

Some of the most powerful and meaningful music written for me was composed by the great Japanese composer Toru Takemitsu. His compositions resonate through nature—its energy and miraculous beauty provided Toru with inspiration. Emerging from darkness underneath the ominous roll of the bass drum, the clarinet releases a multiphonic cry, and as the stage lights come up, a brass chord glistens—Toru Takemitsu's *Waves* has begun.

When composing *Waterway* (written for Peter Serkin and the Tashi group) he watched a nearby stream from his small studio in the countryside. Individual rivulets of water gradually joined together for a time and then separated to move off on their own. In bringing this imagery to life, Takemitsu gave sensitive expression to his good friend John Cage's

aleatoric esthetic: floating among freely played vibraphone, piano and harp gestures are three pitches repeated indefinitely by the violin, cello and clarinet at different speeds, until at some undetermined moment they coalesce into one single melodic stream.

Toru said, "I would like to write music as strong as silence," and I would like to add that the miracle of music in the clarinet is creating a flow of breath through that silence into a wave of sound which sends the tone to the soul, like the breath of a cloud releasing raindrops to caress a flower.

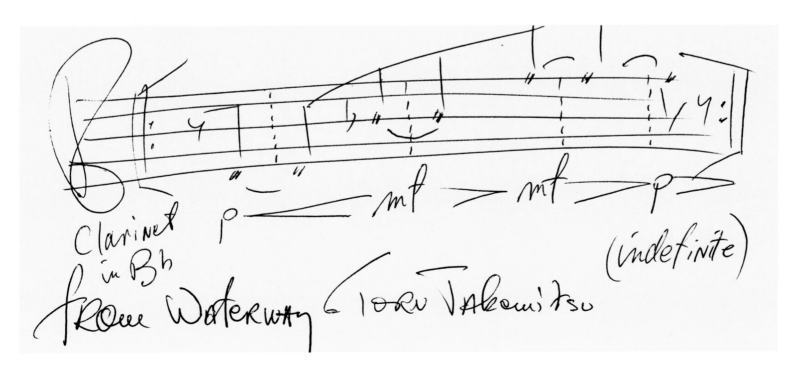

Sketch by Richard Stoltzman.

EVELYN GLENNIE

Water flows.   As do the notes.

From high to low   Across the staves

And through us all,   And through our souls,

Then around again.

WESLEY JEFFERSON

About the water, it's a powerful thing, I'll tell you that. When you go down to it, down to the river, the water is running and you can't help but just concentrate with the river, you know. Looks like it clears your mind, kind of thing, and you just can think freely. It's kind of like a history, so much has happened there. And when you touch that water you're getting a kind of connection to *everything* that happened, sunken vessels and all that, whatever have you there.

Being a blues person you might come up with something that helps you to focus on some of these blues things. They got something in common, a history. They both have the same thing. What done happened on the river and what done happened to you is almost the same thing. So you come up with a song about it, the Mississippi River.

I've been to the ocean one time, the Atlantic, and it had me down on my knees. It was *so powerful*. You feel like there ain't nothing you can do with that but just get on your knees and pray. It's so big.

So, it's pleasure, water is. It's everything. I've been on the Mississippi River in St. Louis. They was playing a little jazz. I got a few ideas from it, 'cause what they mostly done is the blues on them steamboats and gambling boats. I had a great adventure on that. A sister-in-law who was a cook. She got me on there.

The Mississippi is just the largest body of water you can imagine. It has some effect on you when you get down to it, when you go to it and see all of that water. You might be thinking about nothing and that river speaks to you, you know. You probably come up with a song. It acts like it's got control to make you feel closer to the Lord. Like you want to get on your knees and pray to it before you get on it. You feel the more you'll be in His hands. You've done as

much as you can do, you're on your own. The Mississippi has all of that history, all of that death, and all of it connects, and sometimes you can feel all of that, and you're in His hands. It's kind of like space, that river. Ain't nothing you can do out there, when you're out there in the middle of it. It's been a lot of accidents on that river. It just connects with the oceans and stuff. And sometimes you just feel all of that automatic without even thinking about it. You get to staring at it and all kinds of things just be going through your mind. So I believe that's the connection. There are a lot of peoples buried in that water.

Now that levee is real. It's real, man. Every time I cross it I feel what it took back in them days to build it. People lost their lives. We might be driving over some of them that built it. The river has a deep connection to all the things that done happened. They're all kind of related, you know what I'm saying?

When you're on the river, you feel like you're in church. You kind of feel like you can be heard, if you get on your knees. I have prayed out there, out fishing. You just feel like you can be heard when you pray on the river. It can be silent, so everything is heard.

Most blues singers, they was deep in church before they started singing the blues. Just making their living, really. Of course, back in them days it wasn't about 15 cents an hour picking cotton, but look what you could buy with 15 cents. I could feed two peoples. Back in about '66. You could feed a whole family off of 15 dollars a week. And that's about all I would make sometime. Three dollars a day, from sunup to sundown. And I started driving tractors. I come right up through the steps. I was right at the end of the Depression, cotton picking, and then came the machine.

So I would get my fishing pole on a Saturday when I *did*

have some time off, and go to the river. I used to go to a lake called Swan Lake. I was scared, but still I would go down to the water like I had some kind of protection. It's a lot of things that done happened on that lake. It eventually dumps into the Mississippi River. I was down there one day, fishing, and honest to God, I don't know what that was. It was something big come up in the center. Man I ran so fast. Whatever it was shot along fast and went into the bank. They say it wasn't no whale, but I sure believe it was a whale.

Late at night when the river used to flood people used to cut across there in small boats. I remember seven peoples had a wreck out there, and they had to swim. They were all saved. They all had on life jackets and they come up to our house. We were staying way back in the back. They was scared and cold. And we gave them food and blankets. Momma, she used to make quilts *that* thick, man.

The Mississippi, the Sunflower, Swan Lake, all the water, really, connects with the blues. It seems like whenever you get close to a big body of water it's trying to tell you what all went on. You can feel—it's a strange kind of feeling—but you can really feel it. I do every time. I don't know if it's true or not, but it seems like if I look at that water long enough, when it's running and stuff, look like it'll be taking control of your brain, something be going with it. Like hypnotize you. I don't know why I said that, 'cause I believe God will be with you equal everywhere, but it seems like worship is a little truer on the water. I would go down there alone sometimes and just say a prayer. It was like you was being heard. That's what water brings back to me, it brings back what used to happen.

I don't think about it much after, unless I'm at the water, but every time I cross the cold water I think about all of those soldiers going down through the Yazoo Pass, the

Union Army. I be trying to picture what they looked like. I be trying to see what it looked like and what went on.

Thunderstorms, lightning storms. When it started that, I would be quiet. My granddaddy and me, we used to go sit in the corner of the house, turn all the lights off. And I be right there on my knees and he be tuning the acoustic guitar. He be singing blues. He'd act like he couldn't make music 'til it rain. In high waters and all of that. Well, he would say "Lord have mercy" first, and then start mumbling some kind of a song. It was really the blues. Most of it was about hard working. He'd sing a couple songs about women, something like "you done left me alone . . ." or "I wanna leave *you,* treat me like you do . . ." I was going on seven years old. Everybody be in the house, but he be in one part, me and him, and he would be in the corner, and me on my knees, and everybody would listen. Singin' them ol' blues songs. That's when he would really get into it. He could communicate more when it was raining. It was something.

I was about 18 before I knew what blues was. I used to be singing it all of the time. I used to sing in church. I didn't know what it was 'til I got out and met peoples and they told me it was blues. I didn't know what it was. I know I was born singing them. It was like a singing prayer. Because I be asking the Lord to make it better for me, and *get me out of this.* It be like a singing prayer. "Oh, if I ever . . . leave here, what am I gonna do . . ." and "what I ain't gonna do." It don't work like that, but sometimes you have to be doing some of the same things until the process moves, step by step, you know.

I never experienced the flood, but I heard about it. We used to have small little ol' floods; it might take the town or something. I seen some pretty good floods between Marks and Batesville. That big flood makes you feel so strong because it's giving you survival strength. Like you're tested. And you get extra strength, and you start to *sing,* and then you won't be payin' the trouble no attention. Reactin' to what you're seeing, and the others will do it right behind you. It's lifting to their spirit. It's like there's some kind of connection. The other person will be doing just as good as you if he be hearing you singing, such as if you be at work, sand-baggin' and all of that stuff.

Glendora, Mississippi. I was down in there at the time Emmett Till was murdered. Drowned. We was scared. He was cut up first with a knife. We was scared. We see those people coming and we hid. We'd run out across the field and hide. Shooting and stuff. Man, people was scared to death. That's just the way it was. Like I used to go in a store to buy something like tobacco, Prince Albert, or Co' Cola. We couldn't buy a Co' Cola. Colored couldn't buy no Co' Cola. But we could buy soda water. But if you bought Prince Albert, you had to ask for "*Mister* Prince Albert." At that time you couldn't look at a white person. Not seriously. You couldn't look them in the eye. You couldn't look at a white woman, *period.*

The bossman had a spot in the barn. You didn't do right he'd take you there, and straighten you out. Make you get on your knees. He had a long whip. And he'd whoop you. You'd never do it again. I heard that happen, man, someone a-hollerin' and all. I hate to think about it, but that's the way it was done.

That's why I say, "Lord, you done did your work," 'cause I can see all of the changes that have happened since then. You have to know the history to appreciate what you've got now. The world has come a long ways since I was born.

When I take people to the river with the Quapaw Canoe Company, or pick them up, I get there early and try to sing a couple numbers of blues. I don't know why. Like I said, that water just . . . I don't know . . . make you feel distance. And then you feel like you're closer to the Lord, and it's like He can hear you more. Gets you in the feeling. That's where all of the blues comes from, off the rivers and close to the banks, near all of that water.

They did a lot of cotton loading, you know. Take five days to get there, to the levee. They sit around there waiting for it to be loaded, and they're singing and going on. Rhythm and stuff. You had a lot of singers on the steamboats that hauled the cotton. Good blues singers. They would *always* talk about New Orleans. Matter of fact, up and down the whole river. Connections, like at the loading docks, where they could have conversations.

I was always around older peoples. I'd go visit an old gentleman and he'd tell me all these stories, you know? Well, he'd tell me about what happened. Some of them I've forgotten, some of them I haven't. I'd sit up and listen to it. I mean I would be still. You wouldn't get no trouble out of me. I'd be listening to every word. As things happened, I'd growed on and growed on, it just kind of wiped out. 'Cause you know, it's kind of hard to find an old guy now to talk like it was say back in '55 or '60. But they'd talk to me about hard times it was. A lot of them worked on the river, on the water. I know some artists who work on the river now. They're not recording or anything. They be talking about the water, what they remembered be happened. I used to beg Momma to go to one of them's house just to hear one of those men play. He bring it to the river sometimes. They be playing sometimes and they have to make me leave. But he loved to play. We'd just sit there. He'd have lamplights and a kerosene heater.

I lived in Memphis, in '67. That's when "Muddy" Island wasn't nothin' but a muddy island. Just what it says. I used to take Joey, Ronnie and Lisa and them out there to play. And we'd be throwin' stones in the water. That's when I first started playing guitar. I ran off from the boss-man at Stumpy Dead End. I was going on 19. I went on and told a few stories about my age and got on at Anderson Tulley. It was right there at Muddy Island. A logging operation. I was in the furniture side. We'd shape up and make things, like coffins, meat blocks, and made parts for other factories. They'd cut the lumber on one side of the street then ship it across the street to us to finish. I saw many a log ripped open. I used to watch them float down the river. They bunked them somewhere upstream, Arkansas, and then floated them into Memphis, into Anderson Tulley. There was a little ol' dirt road went right alongside the plant, right there on Muddy Island. That's where I'd take the kids every Sunday and let them play. They wasn't but three years old. The weeds was *tall* and everything. And I had me a little 'coustic guitar, and I'd sit down and make it play.

The river gave me the same feeling up there it did here. It still hit me, the history thing, what happened. You might be thinking about what you might do, and by the river you come up with some winners. But you be kind of silent. The river is kind of a silent thing. People on the river, they don't say much. I don't know if they be more careful, or what. You *know* the feelings you get.

Tornadoes. I think everybody'd bow down when a tornado's coming. I remember one night. I's just got married. I never seen a tornado before, I never heard it, but I heard

something just going "whrrrrrrrr . . ." just like six or seven diesel trains. And my sister, she was living with me. My wife was just about 13, and I was 18. The judge let us get married because we had went together and they thought she had gotten pregnant. But we never done shit . . . it was a year before we'd done anything, after we'd got married.

My wife was scared. I said, "Y'all, this storm's fixin' to hit this house." Say from one side of the street to the other, that's how close the tornado passed the house. It just cut a *path*. And I looked out. I just had to *see*. And all that big black . . . it looked like smoke. It be black as the devil. And it just tore into those big cottonwood trees. It cut a path about a half-mile long. And hail. Big hail. That's the reason I knowed it was something. And it had hail look like as big as a baseball. It started dumping that before it got there. When it got closer, the more it fell. It beat that little tin-top house. And those beds was *long*. We had heard a little bit about what to do in a storm. We was so fat, and those beds was about so high. We done got our head in there and our ass out! I said we done no good. We're worse than a stork, or something, hide its head and all this ass is out! Boy, we talk about that sometimes now. We was some kind of scared people.

But the Lord, He done bless me. I don't know. A lot of my friends, they done gone on. But I must have done something right. I'm still here, you know. All those people I know who would have been the same age as I, is done gone on. I tell my wife Sarah, it must have been something I was doing right. I know He ain't ready, 'cause when He get ready, God's gonna get you. Maybe there's something He wants me to do or see. I always try to *help* people a lot. I was the kind of person that people'd come to and get advice. And I'd give them the right kind.

KENNY LOGGINS & JIM MESSINA

PAULA ROBISON

A Dialogue

The listener says

　　—I'm thirsty . . .

The flute player says

　　—Well, I'll pour out my song for you!
　　I'll sing of the earliest light on the earliest
　　river, the first bird call and an answer,
　　the first step of the dance and the answering
　　step, the joining of hands, the mysteries—

　　I am the song of all these things, you know . . .

The listener says

　　—But where is the river? Where are the
　　deep places and the trees heavy with fruit?

The flute player says

　　—I am the source: the sparkling spring!
　　I caress you. I dance the first steps
　　with you. I flow to the stream with you.

　　—As for the river, you'll have to find
　　it for yourself.

　　—When you get there, remember me . . .

BOB DOROUGH

Arkansas was laid down upon this Earth to give us

strength and depth and moisture.

The valley of the vapors—mountains, tall and strong.

     Nowadays folks go crashin' through this wilderness we

call "civilization"—we didn't do that, naw, not in Arkansas.

In Arkansas, everybody moves around kinda carefully,

They tiptoe through the bedroom and close the door behind so quietly.

In Arkansas folks don't go makin' lots o' noise, they don't go

pickin' up your toys without askin'—not in Arkansas.

In Arkansas, oh the water is plentiful, gushin', it's fallin' and

rushin', bubblin' and slushin', fillin' rills and rivulets,

polishin' stones—beautiful, flashing waters.

It's so sweet there, from all o' the springs and pools it seeps

there, you can drink it right out of the creek there, the waters

of Arkansas.

From "Not in Arkansas," a song by Bob Dorough, from his autobiographical musical, "Four of My Nine Lives." © 1990 Aral Music Co. Used by Permission

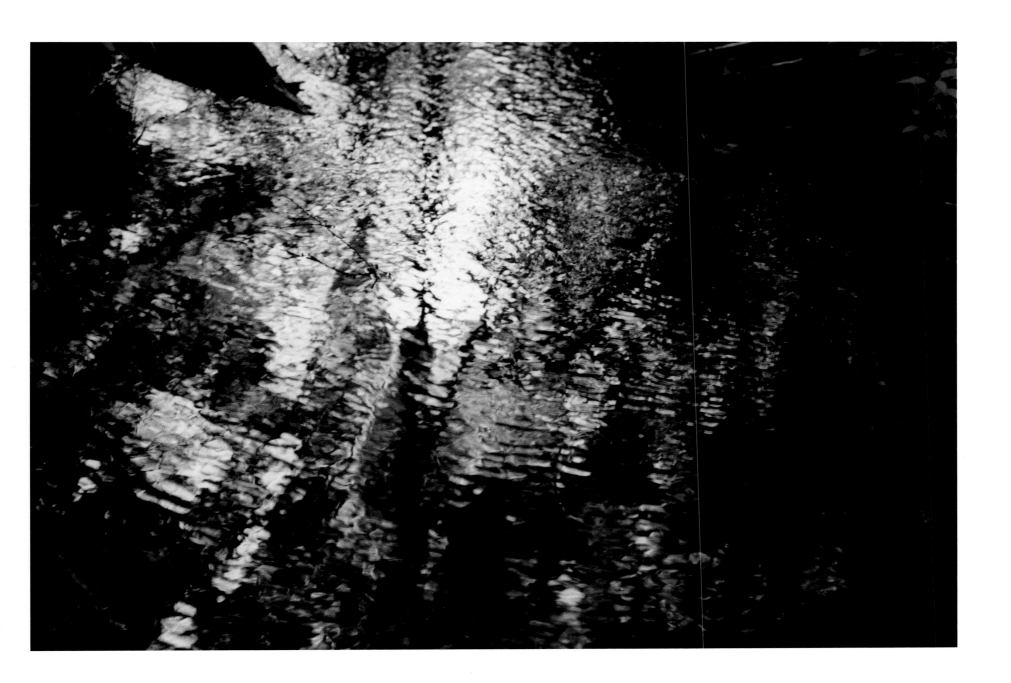

Water, like music, is absolutely essential to life, and their

similarities are almost frightening. Both have an

effortless ability to reduce a person to tears, or to

soothe and bring serenity to one's consciousness. Both

are also capable of simultaneously bringing forth a

torrent of turbulent emotions that can lead to a

passionate eruption of almost terrifying strength.

SARAH CHANG

Water and music, quite simply, are two of the most beautiful

and precious gifts in life.

PAAVO JÄRVI

For me, water is much more a mental state than something physical. When I am

beside a massive body of water, it is the only place I feel complete freedom.

MISCHA MAISKY

I have always been strangely drawn by water, and while this fascination could not be compared with my lifelong obsession with music, there is somehow a connection. I was born in Riga on the Baltic Sea, and this was my first exposure to the majesty that a body of water could evoke.

Later, the pursuance of my musical studies led me to St. Petersburg, where both music and water flowed like blood through the veins. When I repatriated to Israel, the warm Mediterranean represented hope in that parched land. Studying in Los Angeles with Piatigorsky was a dream fulfilled which also introduced me to the cold, wild waters of the Pacific.

When I fell in love with Kay, we made our home in Paris overlooking the mysterious and beguiling waters of the Seine. I even fantasized about living in a houseboat in the center of Paris, but this did not seem advisable since I had never learned to swim. Being so attracted to the world's waters spurred me on to correct this weakness, and at the age of 30, I went to a special school where I relearned what a baby knows in the womb.

Now, my family and I live in Belgium, a country of water and rain, on the banks of a small, charming lake teeming with life. We are all drawn to music, which is as essential to us all as the water that makes up 70 percent of the world's surface and 70 percent of our bodies.

We musicians can learn much from watching the water in its calm, flowing form, sometimes followed by turbulence only to search after and find, once again, peace. Water is music. Water is peace. Water is life.

PAMELA FRANK

Water, as music, has always moved me. This reaction, until now, has been visceral and unconscious. But upon further reflection, I realize this is no coincidence: music and water share the same properties and powers.

There are many parallels between these two all-important ingredients in human life. First, the physical: the human body is made up mostly of water, the earth's surface is covered mostly by water, and we come from the protection of water in the womb. Though not as literally, I, as a musician, depend on music for spiritual survival—it is my oxygen. We all know that we cannot live without oxygen and water.

Second, music and water are inextricably linked by their very definitions. Ebb and flow are the common denominators. Water dances and undulates. It is the physical embodiment of the musical term *rubato:* the giving and taking of time. This sense of back and forth and push and pull is what creates the feeling of aliveness and life. No matter how still a body of water is, it is always moving. There is no such thing as stasis. This constant, inherent motion is also the essence of music. It vibrates because it always comes from something and is going somewhere.

Personally, I am attracted to water for the same reasons that I am drawn to music. It has a complex, natural character: It can evoke stillness and tranquility (providing a calming effect) as well as turbulence and volatility. It can soothe as well as thrash and crash. As such, it elicits an emotional, gut reaction—that is the purpose of music.

When looking to nature for solace or inspiration, I've never asked the mountains for

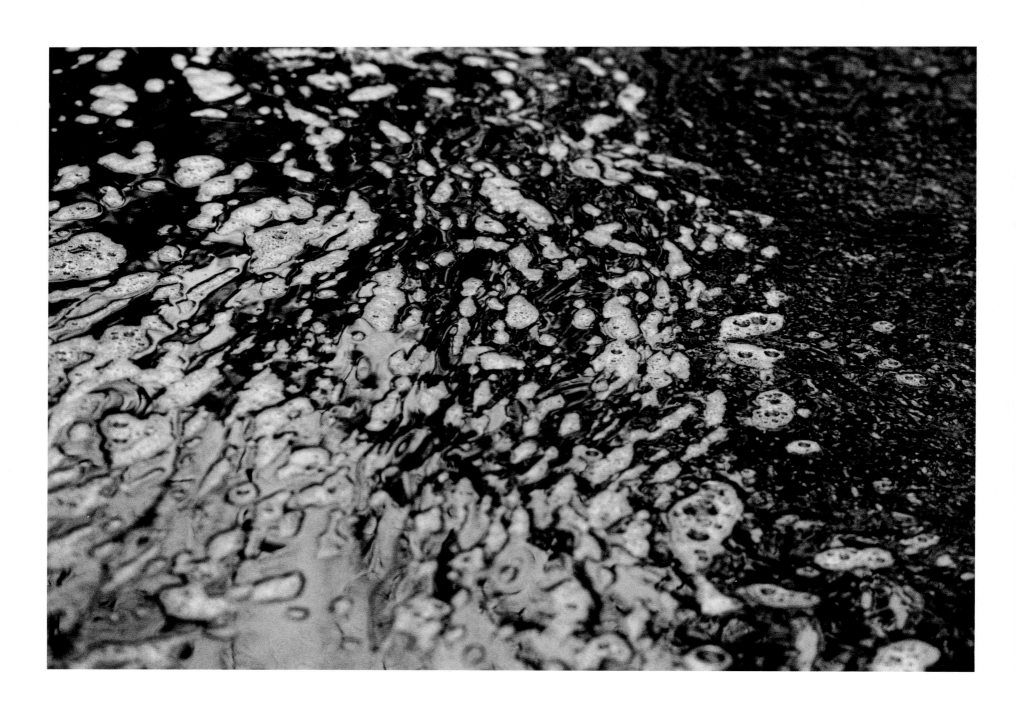

guidance—they are too concrete. The peak is obvious, the beginning and end of the journey far too clear and definable.

Water, however, is infinite. I get lost in its vastness. It is larger than life and can surround and engulf us and our senses. This provokes my imagination and forces me to question and wonder what is beyond it. That, too, is the essence of music and its effect on me.

Water has a sound all its own, and I find comfort in discovering what it tells me. I often try to let the music play me, rather than the other way around. And so, both inspire me to listen.

In the end, music and water are both undeniable forces of nature, and humans are very small in the face of either of them. Both have the power to sweep us up and carry us away, to envelop and overwhelm us. The *only* difference lies in the consequences thereof: To be drowned in water is not enviable. To be drowned in music is.

ERIK NIELSEN

Purling

I chose the title for this little composition because it expressed better than any other word the incessant and ever-changing motion of water in a stream, which I was trying to convey musically. As I wrote the piece and later, when I struggled to find the right title, I thought of the stream near my house when I was a boy. For several years it was an integral part of my life, as I swam, fished and even skated occasionally in or on it. I also walked along it times without number, observing it through the changing weather, seasons and natural conditions that made it always the same and always different. The quality I remember most was the motion: circular, up and down, straight ahead, all simultaneous, thus making a complexity of separate shapes and directions which somehow made a whole with no beginning and no ending. Though I have not stood by the stream in about 20 years, I suspect it still moves in the very same way. I can see it clearly.

MIKE GORDON

Have you ever visited your Mother's astrologer? I did. And I have to admit that her voice sounded soothing and inevitable in that candle-lit sitting room. Sometime in the middle of reading my chart and tracing my past lives, she issued an alluring little nugget of insight: that for me in particular, the way that I can predict the future, or meditate, or find my spirit, is to "watch water." It just so happens that I have a window to the cosmos, and that glassy, rippley surface awaits my gaze.

The concept resonated, but I should go back and examine my history with water. At age two, I was accidentally dropped into a swimming pool, but had fun being rescued.

At six, at Camp Elbenobscot, I failed all swimming lessons. The coldness of that pond made me too nervous to breathe.

At eight, I designed and built, with the help of my Father, a paddleboat. We brought it down to the Sudbury River, and with all of my friends watching, I tested my new contraption. The bicycle-driven paddle mechanism actually worked fine, but the boat sank since the body itself was just a flat piece of plywood.

I was fourteen and standing in a Bahamas swimming pool when I chose my career as a bass player. My Dad and I were watching the Mustangs play at poolside, and I told him if I ever were to join a band, I'd want to play bass because the notes felt good.

At sixteen I began my first major accomplishment by dragging out a garden hose and flooding and digging on a hillside by my house. This would be the foundation for a little cabin, which all of my friends helped build.

At seventeen, after looking at twelve colleges, I chose the University of Vermont for one main reason: I liked the lake. There I met my band, and one thing they all did after band practice was to fly high above the Williston Quarry on a rope-swing. Being afraid of heights, I avoided this activity for a year or so. But I decided that if I couldn't overcome this fear, I wouldn't be able to overcome the fears that would accompany my career as musician, so when no one was around I swung out and dropped the thirty feet. I landed on my face, but I felt good about it.

So what do I do with this gift I have? Pour myself some water, stare at it, and solve all my problems? Go to places like Niagara Falls and bring some troubled souls along that need healing?

Actually, from time to time I find myself walking and pondering, and inevitably there is a lake, river, or ocean to peer at from a bridge or sandy shore. So I stop and gaze. The rippling seems random and mesmerizing. Every point in space, each spot on that meniscus where my eyes might focus, flickers with all shades of blue, green, black, brown, and beige. The flickering is instantaneous and constant. The experience is at once cathartic and chaotic. Enveloped in entropy and organized by order. Yeah, she was right. That's my portal.

I have not even a smidgen of a clue to know how the water is helping me. I learn nothing, feel no wisdom, gain no insight, but I simply trust that my soul has satiated its thirst in some unknown way, and I walk away a better person.

MIDORI

Water is a substance of contradictions.

It has fascinated and enchanted us; it has annoyed us.

When was the last time a favorite book

left in the garden was soaked?

Remember that summer day, swimming in the lake?

The morning's first sip?

Yet, water has the forbidding power to destroy and kill.

Like fire, it is a thoughtless, splendid thing,

perhaps better to be left alone than to be mastered.

Like the world, it is always in motion, flowing toward some place:

constant and unstoppable.

WILLIAM PARKER

Overcoat in the River

Water bird sings
netted swallow rested on bright cloud
sun water, laughing against
blue, black, purple water cathedrals
flowing gardens illuminate miles and
miles of unpicked flowers
they float on this golden, green river.
Three men wearing pointed bamboo hats
fish on its shore
they cry out
"Freedom for all human beings."
"Freedom for all human beings."

This the painter's dream
portrait of the root world
frogs and snakes, worms, turtles and crickets
in this place I would listen to Uncle
Marvin play what he called water music,
from his tenor saxophone
now if I were standing one inch from
the bell of the saxophone
I could not hear the sound
that's how deep this was
Uncle Marvin would always
wear a long black overcoat
winter or summer spring or fall
he'd tap his foot and do a little dance.
inside his shoe his foot was blue
and it kept time to all the changes
that would bring us back to the beginning.

water song, water sound
painter's dream is to see
to feel to understand
that someone cried a lot of tears
to make all these oceans, rivers
and lakes
now whether they were tears of joy
or tears of pain I don't know
Uncle Marvin continued to play
his horn and I continued not to hear
and the river continued to move.
On the last Saturday morning of the year
I sat at the left arm of the totem pole tree
at the edge of the river.

A black overcoat is in the water—
I hear sounds that could only
be water music
sounds that could not be repeated
they are followed by raindrops
then sunshine
a double rainbow appears
compassion for all human beings
overcoat in the river
overcoat in the river
inverted flower equal child
water song, water sound
painter's dream
world not die
just beginning to live
song of the water
sound of the water.

GREGORY TURAY

Eternal Water

The first time I ever saw the ocean I was awestruck at its incredible presence and size. I had never seen anything so enormous and powerful; at the same time, I felt so small and insignificant. The second thing that struck me was the incessant rolling waves beating on the shore just as our hearts incessantly beat in our own bodies.

Those two things in themselves were enough to leave quite an impression, but there was yet something else. The last impression this magnificent body of water left on me was nothing more than a question. It was as if the ocean whispered to me, asking me, "Who are *you*?"

"Who *am* I?" I thought to myself. "Here I stand at the foot of history itself. Where do *I* come in? Where and how does the river of my life flow into this 'history'?"

I began to sense something truly greater than myself, even greater than the ocean itself, as powerful as it was. In that moment, I came face to face with my Creator. The very Creator of all the earth, including that ocean. It was God, the Eternal Spring from which all life and water come. Knowing this and knowing Him, I have come to know my true existence and meaning. I have found the True Water of life. (John 4:13,14)

RANDY WESTON

Water means so much to me.

After a very serious attack of arthritis in the early 1960s, I was cured by an 80-year-old doctor with aspirin, rest, vegetables, massage, and hot water. He told me that hot water was my savior and that whenever I got wet or caught in a draft, I should use water hot in the form of a shower or bath. A few years later, I was living in Morocco. There I discovered the ancient tradition of *hamam,* or community baths. For thousands of years, the African people have replenished body and spirit by immersing themselves in hot water. The *hamam* were even brought to Spain from Africa. In the first room, an incredible heat overwhelms the body, followed by a submerging in water. This hot room leads to a second, warm room. The masseuse chooses one of the rooms for a very spiritual cleansing process: the body is laid down on heated tiles, massaged, covered with soap, and whisked entirely with a harsh brush. Then the body is rinsed with cool, clean water, and all the dirt is shed. The third room is refreshingly chilly and quiet. Here you sit on mats, wrapped in towels, and sip water in the form of hot tea, or perhaps orange juice. You take time to reflect and meditate in the silence. Everything else seems to just stop. I felt close to God in the stillness of that cool mist.

That doctor's words of wisdom about water saved my life. The rituals of water and massage at the *hamam* renewed my spirit.

When I saw the Nile in the late 1990s, I felt I was witnessing the greatest water on earth. I had read about the Nile River and Nubia since I was 17. History comes from that river. It's where it all began. It's the birthplace of great civilization—the birthplace of Egypt and Nubia. You can actually see that miraculous birthplace from the water.

For that Nile trip, I was with my lady friend from Senegal. It was the first time for both of us on that great river. I found myself deeply moved by the magic of the river, by its beauty. It was very powerful; it was transporting. We were so blessed to be on that river, where it all began. I composed my jazz piece "Ancient Future" in response to that spiritual journey down the Nile. In all my life I'd never felt anything like it.

I had a recording with me of the music of the Master Gnawa Musicians of Morocco. Their music is so old, not even the Master Gnawa understand all the words. It is truly the ancient music of Africa. I played it for all the Egyptians on the boat, including the captain and his crew. When I first saw the captain I expected someone typical, dressed in white, like the captain of a ship. But here he was in his traditional robe, all laid back. People told me that the captain knew the Nile so well that he could go to sleep and still take the boat down the river. None of the Egyptians on the boat had ever heard the music of the Gnawa before, the music of their ancestors. They had never heard anything like it. But floating down the river that day, they heard their own ancestral music for the first time. We all share this ancient music. The Nile, birth water for civilization, was a natural and powerful place for the Egyptians to make the discovery.

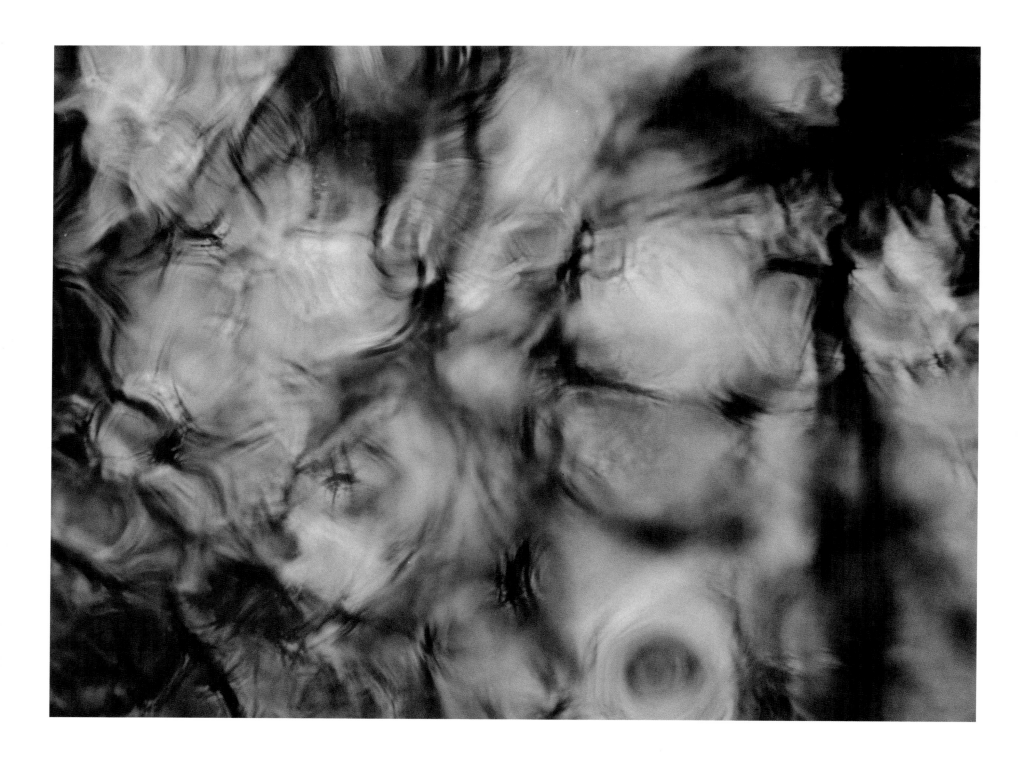

SUSANNE MENTZER

When I was twelve my father was the superintendent of Catoctin Mountain Park (a national park) in the mountains of central Maryland. It was an idyllic life, in the woods with only one house nearby, miles of hiking trails, streams and wildlife, whippoorwills giving deafening concerts in the evenings. Aside from the occasional interruption of President Nixon's entourage heading up to Camp David at the top of Catoctin Mountain, it was a lazy, bucolic setting.

Across the road from our house was Whiskey Run, probably so named for the illegal stills that used it to cool various concoctions in the first half of the twentieth century.

I spent many hours in this shaded stream on hot summer days, looking for crayfish, building little dams, and watching little stick boats that I made float down over the rocks through elfin-size rapids.

But my best memory has to do with my discovery that I had a voice. I had a wonderful friend from school who sang soprano in the school chorus. I sang alto. Denise came to visit one stiflingly hot day and we, not having much to do in our adolescent summer boredom, headed down to the creek. There was a culvert under a road through which Whiskey Run traveled, and so inviting was it that we climbed into this galvanized metal tunnel with our clothes on, sat curled to the shape of the tube and let the water flow over us as we sang, in harmony, just about every song we knew and some that we didn't.

We must have stayed in that echoing chamber for hours, where no one could find us, our voices amplified by the cylindrical chamber, and we got lost in the bloom of discovery of music making, accompanied by babbling Whiskey Run.

SHARON ROBINSON & JAIME LAREDO

A Toast

Were it not for the magic of water

There'd be no long soothing soaks after a concert

No falling asleep listening to the gentle rain

No witnessing of a majestic waterfall

No rejuvenating water massages

No viewing the tugboats on the mighty Hudson

No joyous swimming at Miami Beach

No luxurious sailing on the seven seas

No lazy sunsets listening to the waves

No snorkeling at the Great Barrier Reef

And worst of all—nothing to splash in our Scotch!

So here's to Mom,

Mother Nature, that is!

She nourishes us, body and soul,

With that precious liquid gold—WATER.

MARCUS ROBERTS

I respect the water.
Even when you can't see it or touch it, you can hear it.
Even if you can't hear it, you can imagine its presence.
In a glass, in a bowl, in the sink, on the floor, from heaven above, in a lake, stream, pool, or ocean.

I respect the water.
Its soothing clean flavor protects and nourishes.
Its cool texture rejuvenates, gives newfound strength to tired muscles and bones.
We compete on and under its surface, play in it, revel in its power and beauty.

I respect the water.
Because it gives life in one moment and can take it in the next.
It is capable of causing tremendous pain, suffering, and harm.
It can rip apart cities, destroy entire families, return buildings and trees and people to dust.

I respect the water.
Because it's been here forever—feeding nature, quenching thirst, healing the land, nurturing the flowers.
And it will be here long after we are gone, demanding and receiving respect.
Singing the sweet song of eternal life and playing a groove that all cultures can feel.

I respect the water.
Its taste, texture, touch, sound, and image a reflection of every living creature on earth.
As timeless as a folk theme whose origin no one recalls.
It gives its melody to all generations, showering us with abstract dissonance and consonant beauty.

I respect the water.

MICKEY HART

I have always had a certain fascination with rain. When I was a small boy, I always liked to go out in the fiercest rainstorm and stand there under its spell. The drops hammered away at my consciousness, transforming me into some kind of drum. When my Mom saw this happen, she would try to coax me back into our house to no avail. She once asked me what it was about the rain that so attracted me. I told her it was like music to me. What I was experiencing at that time I now know to have been an epiphany, a moment of extreme ecstasy, and a rapture that resonated deep in my subconscious. The rhythm of the rain was my friend, and nothing was more soothing and trance-like. In my young life, it felt like water from the heavens that was speaking directly to my soul. As I look back, I realize that I have used the sound of the rain in so many of my recordings. Until now, I have never really paid much attention to this strange allure that was born in my youth.

TILLA HENKINS

An enthusiastic farmer and guide for our orchestra when we were on tour in Zululand surprised us all when he told us that they actually do not welcome rain in that area: "It is the early-morning dew caught and absorbed by the pineapple leaves that is actually the key to the sweetness of our fruit!"

Upon his remark that the people of his region were about to experience their first live symphony concert since the first pineapples were planted there 25 years ago, my mind drifted back two years, to another seemingly wild, unspoiled woodland, home of the Interlochen Arts Camp in Michigan. Forty trombones were sending ethereal-sounding notes through the early-morning mists across Duck Lake's waters into the woods where I was strolling, leaving me totally spellbound.

It was also at Interlochen that I met a young girl from Israel who surprised and delighted me simply by sharing my name.

"How did you get your name?" I asked, not mentioning the unglamorous status of this rare nickname back home in South Africa.

"Well, when I was born," she proudly answered, "my mother wanted to find the most beautiful, unique name she possibly could for me, one with a life-giving meaning. Did you know that 'Tilla' means 'dew'?"

It was worth going halfway around the globe to discover that I had a noble name after all, one that represents the morning dew that brings sweetness to life.

SHARON ISBIN

Beginning in the late 1980s, I made the first of several trips to visit rain forests
in Costa Rica and Ecuador. Floating down the Amazon in Ecuador, in a dugout canoe
with piranhas, electric eels, and glistening crocodiles afoot, monkeys, sloths,
toucans, macaws, and an occasional python in the lush foliage overhead, I was in
a state of bliss. Surely, this was the Garden of Eden. In this world, water was the magical
conduit of life and its ancient origins.

In the Galapagos, I traveled by boat from island to island, witnessing
the majesty of destiny. These were the waters of Charles Darwin, and the
extraordinary life forms they nourished, from sea lions and iguanas to exotic frigate birds and red-footed
boobies. In Tortuguero, Costa Rica, I watched huge leatherback
turtles arduously make their way from the safety of the ocean's edge to the beach
where they would lay their astonishingly beautiful spherical eggs.

I had no idea then that these experiences would come to figure in my music as well.
Back in New York, I met the composer/percussionist Thiago de Mello, an Indian
from the Maue tribe of the Brazilian Amazon. He shared with me his compositions which describe the
legends of his people, and evoke the sounds of the Amazon
and its inhabitants. We began to perform together as a guitar and percussion duo,
and eventually our collaboration led us to make our Grammy-nominated CD,
Journey to the Amazon, a collection of Latin American music from countries of the Amazon and its tribu-
taries: Brazil, Paraguay, Venezuela, and Colombia.

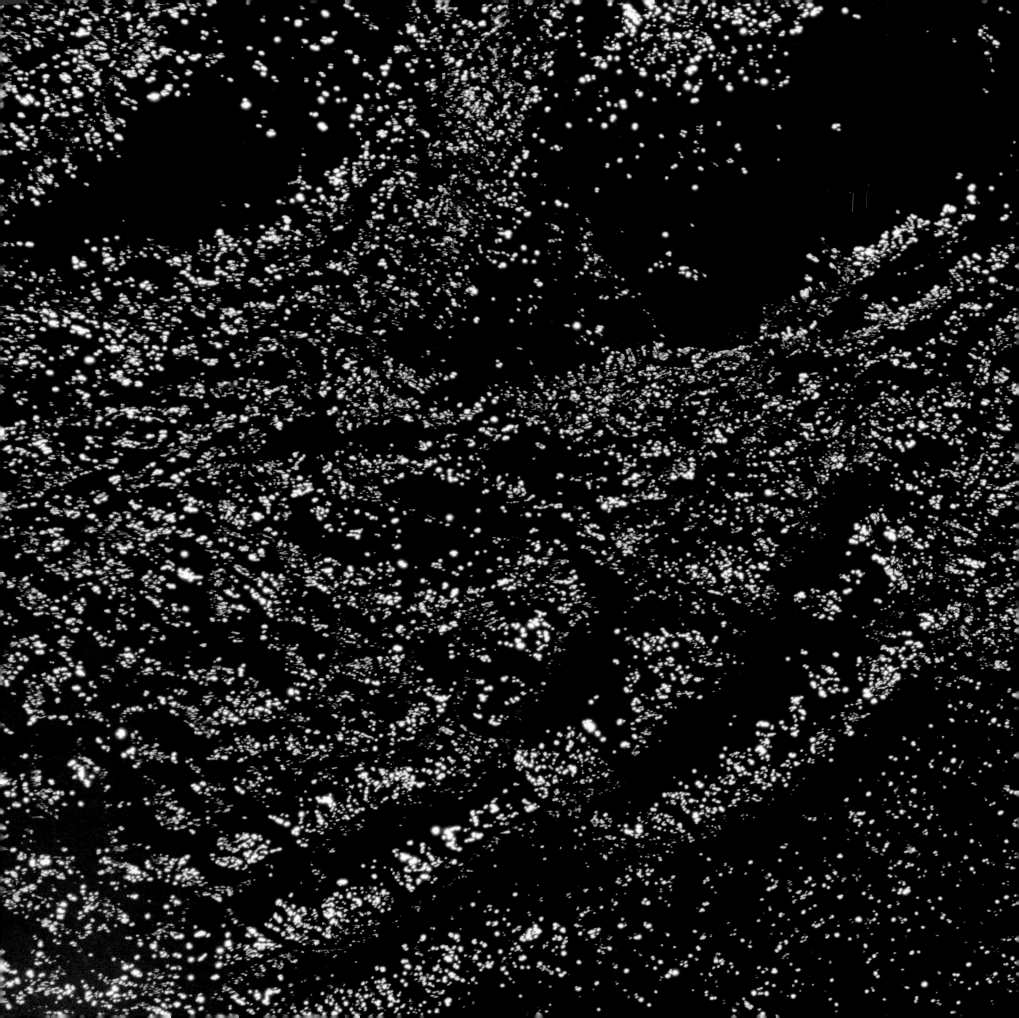

CHRISTOPHER PARKENING

Even before I began playing the guitar, I had a great love of the outdoors and rivers, in particular fly-fishing for trout. My dad taught me the art of dry fly-fishing when I was six years old, and the most enjoyable times of my life were spent on a trout stream in the High Sierras of Northern California. My goal in life was to retire at age 30 and own my own ranch with my own private trout stream.

At age 19, I signed a contract for a series of six albums, and was asked to start a guitar department at the University of Southern California. The following year I embarked on a rigorous concert schedule touring the world, eventually performing over 90 concerts a year.

Needless to say, as I added a grueling concert schedule to my teaching and recording obligations, my life became ever more stressful. Frankly, I was miserable on tour. I hated the hotel rooms, the airplanes, the monotony of one concert after the next. However, this exhausting schedule allowed me to attain my goal—at age 30, I stopped playing the guitar, found a ranch with a beautiful trout stream in Montana, and I moved there from Southern California. For the next four years I was doing everything I wanted to do. I was fishing to my heart's content, learning every trout stream in the area, and going back to Southern California in the winter to escape the snow and cold weather. I was living the good life—or so I thought.

One day, a friend invited me to church, and I heard a sermon based on the book of Matthew, chapter 7, verses 21–23. I knew that passage applied to me! When I stood in judgment before Christ, He would say to me, "You never cared to glorify Me with your life or with your music. All you cared about were your ranches and your trout streams. Depart from Me, I never knew you!" It was in that sudden, terrible moment I realized that I was not a Christian. Knowing I was a sinner before God, I prayed and asked Him to forgive me. It was then that I asked Jesus Christ to come into my life, to be my Lord and Savior. For the first time, I remember telling Him, "Whatever You want me to do with my life, Lord, I'll do it."

My new commitment to Christ gave me a great desire to read the Bible and learn more about the Word of God. One day I read a passage from 1 Corinthians which said, "Whatsoever ye do, do all to the glory of God" (1 Corinthians 10:31). Well, there were only two things I knew how to do: one was fly-fishing for trout, and the other was playing the guitar. The latter seemed the better option to pursue. So I sold my ranch and trout stream in Montana, returned to Southern California, and began playing and recording, but for a new purpose—to honor and glorify my Lord and Savior Jesus Christ.

I still have a love for rivers, trout streams, the mountains, and especially fly-fishing, but I now have a greater appreciation for the beauty of God's creation, because of my faith in Jesus Christ.

YOUSUKE MIURA

When I was a boy, being unable to bear the sadness of having lost my grandmother, I ran out of my house and, standing by the brook, shed many tears. The stream was flowing just as it always did, as if nothing had happened.

My country of Japan is blessed to have a lot of water both inside our country and around our island. Most of the rivers are very rapid streams that run from the mountains to the ocean. Our rivers are not wide and slow-moving like the Mississippi River, the Amazon or other big rivers that run through broad plains. But our small streams provide much beautiful scenery and support the lives of all Japanese people, animals and plants.

Looking at the brook that day, I realized that just like the flow of water, the currents of birth or death in our lives also flow forward. No one can stop that flow. As I watched that clear stream flow forward, I came to understand that water is even capable of washing away our sorrows and the pains in our hearts. It has the power to make our hearts as clear as the water is itself.

EUGENE SKEEF

Ngoma—The Story of Nomvula and Her Sacred Drum

Mantaba, goddess of the rain, lived on top of the highest mountain, from which all the rivers flowed. She was also called Mandaba, because she held in the memory of her dreams all the wisdom and all the stories ever told. Her story is our story.

She had long, long hair that ran down on all sides of the mountain, bringing sweet water to the fertile valleys. She wore a crown of colored beads to keep her hair in place, so that the rivers would know exactly where to flow carrying the stories of life and its secrets.

In one of these valleys, which was called the Valley of a Thousand Songs, lived the peaceful Rainbow People. They enjoyed bathing in the Rainbow River because when the sun was high in the blue sky it reflected the beauty of their different colors. There was a lively young girl among them called Nomvula, who took to the warm water like a fish. Her mother had given her this name because she was born in the time of the heaviest rains, when there was prosperity and happiness all around. She was born with the gift of music. She was a great singer, a great dancer. She inherited the ancient rhythms of the elders and the honor of being the chosen drummer to call the rain. When she played the drum, when she sang, all the Rainbow People joined in and sang and danced with her.

Out of the middle of the Rainbow River grew a sacred tree without leaves. The tree had stood there for as long as the river could remember. Not too far from this tree, at the very bottom of the river, in a cave behind a big boulder, lived Mamlambo the great python. The python kept in its belly the precious stone that held all the stories ever told,

which Mantaba passed down the clear waters of the river.

Now there was nothing at all in the world that could shift the huge boulder that blocked the entrance to the cave. Nothing, that is, except Nomvula's music and dancing when it reached a peak of perfect harmony with the rest of the Rainbow People.

This of course happened every season during the festival to celebrate the coming of the first fruits. Using her drum Nomvula called a procession of drummers, singers, dancers and players of all sorts of instruments. Together they gathered the whole community and led them to the festival arena at the edge of the river. They were going to appeal to Mantaba for rain. The bright afternoon sun tuned the instruments to just the right pitch and reflected the rich colors of the Rainbow People's beautiful costumes. With a voice as sweet as honey and the grace of a gazelle, she sang and danced for Mantaba, her drum hanging from her nimble waist. Gradually the singing filled the atmosphere as the Rainbow People harmonized and formed a dancing circle around Nomvula and the other musicians. The feeling was so powerful as the voices of the people and the instruments rose with the dust kicked up by all the dancing. It went higher and higher until it reached that note of perfect harmony—when all the sounds were as one sound and the dancing seemed to flow without any effort at all.

And there! Right there. When the music hit! When it was perfect. That is when Mantaba was so inspired she shook her hair loose and laughed in a state of pure joy as she flung the colorful beaded crown into the open blue sky. The crown turned instantly into the most beautiful

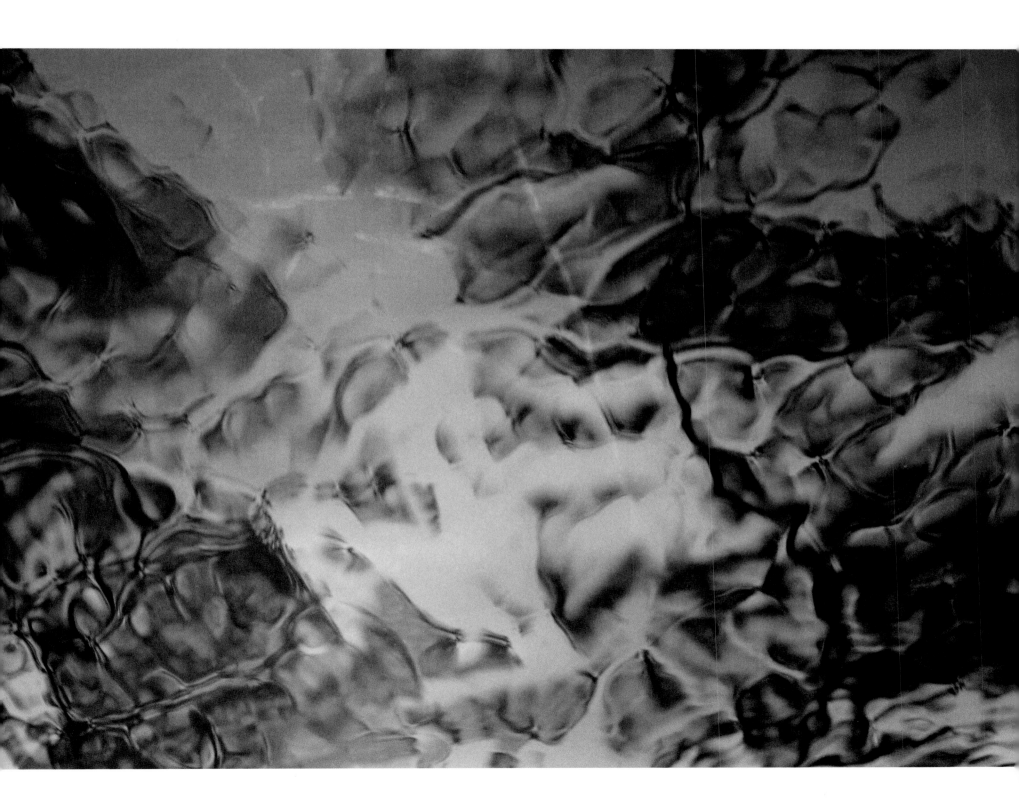

rainbow, which grew out of Mantaba's mind, arched fully like a river of color in the sky, landed over the sacred tree and dipped into the clean water.

And water flowed down into the fertile valleys. And the streams rose up. And at the bottom of the Rainbow River which snaked its way through the Valley of a Thousand Songs, right at the very bottom of the rising river, the boulder rolled away from the cave mouth and Mamlambo swam slowly, slowly out of the cave toward the sacred tree. The python curled its glistening body round and round the tree until it reached the highest branch, and from its stretched belly produced the stone on its tongue. Because in the story stone are all the stories ever told.

Now at this time the music was perfect. So Nomvula could come away from her drum, come away from the dancing, and the music continued even without her. Seeing Mamlambo on the leafless tree with the precious story stone radiant against the rainbow, she dived into the river and swam fast toward it. Propelled by the beautiful rhythmic music of her people, she climbed the tree singing with the most powerful voice. Gently she took the stone from Mamlambo's tongue and held it in her hand. With closed eyes she held it for one whole chorus. And when she did that the knowledge of the stories passed down her arm and into her mind, and into her heart, so that she could tell these stories to the Rainbow People around the night fire for the next year. Then she placed the stone back on Mamlambo's tongue. The python took the stone back into its belly, and then shed its old skin, which it had outgrown. Nomvula was beaming with joy as she pulled the new rainbow from the sky and covered Mamlambo with it.

The Rainbow People danced toward the water's edge to meet Nomvula as she approached, sitting astride Mam-

lambo, with the old skin wrapped 'round her shoulders. The python set her down gracefully and, leaving a shimmery streak in its wake, swam back in its dazzling new coat, down to the cave, where the boulder rolled back in place.

Back at the drum compound the drum makers skinned new drums and repaired old ones with the skin from Mamlambo. And so they were able to play drums for another year. This happened every single year when the rains came.

Now some people were jealous of Nomvula because of her successes and because of her musical abilities. The main one was her stormy older brother. His name was Mkhonto. He had a head shaped like a spear, and a habit of leaning forward when he walked, which made him move faster than most people even when he was not in a hurry. Mkhonto was renowned for his poor hunting skills, like when he was once butted in the buttock by a wild boar he had been tracking for a whole day. However, he was the leader of a gang of evil sorcerers called the Bathakathi. He thought he could do everything better than his sister. He thought he could drum, sing, dance, swim . . . do everything much better.

So he had a go. He gathered some of his friends, and together they tried to make some music to make Mantaba laugh, to get the rivers to flow and to get the cave to open so they could take the power of the story stone for themselves. But it was a complete disaster. When they started singing and drumming everyone just ran away screaming: "Oh please stop! It's appalling . . ."

So the Bathakathi sat around thinking: "What's going wrong? Why doesn't it work? We want the power for ourselves." They scratched their heads in frustration. And as they scratched their heads something strange happened. Large flakes started to fall to the floor from their heads.

One of them noticed that as an ant was coming along the floor a flake fell through the air and landed *qatha!* on its head. The poor ant died. This gave them an idea. They all sat around having more evil thoughts. The poison flakes poured down with a crackling noise as they scratched their untidy heads vigorously (there were about thirteen of them).

It was not long before each of the Bathakathi had a huge pile in front of him. Mkhonto, from behind the biggest one, ordered two of his boys to run to the storage hut and fetch some old potato bags. They filled them with the evil thought flakes and fastened them tightly before hiding them.

Next time the rains were due to come Nomvula brought her drum out and began the ceremony. The Rainbow People joined her, and as Mantaba's brilliant rainbow touched the sacred tree, the rivers rose and the boulder rolled away. The python Mamlambo, in its tight skin, swam slowly out of the cave toward the leafless tree. At exactly that point Mkhonto and the Bathakathi came out of their hiding place carrying the evil thought-flake bags, and emptied them all into the river. With a *poof!* the Rainbow River turned immediately into porridge. Just like that, it turned into this poisonous porridge which was steaming and bubbling and getting thicker. Mamlambo tried to swim further but failed, rolled over and died. Mkhonto took his knife, held it between his teeth and ran. (He nearly fell over his feet trying to keep up with his head.) He swam through the porridge determined to get the power for himself. When he got to the python he took the knife from his teeth, ripped open its stomach and reached for the story stone. But you can't just take a story stone like that. It was much too powerful for him. There was a

huge explosion. He back-flipped through the air and landed unconscious—a sort of belly-flop—in the porridge.

All the music started to become chaotic. It all just fell apart. In the confusion Nomvula could see what had happened. So she ran to the porridgy, poisonous river and dived in holding her breath. She swam all the way down to Mamlambo. She took the python and wrapped it 'round her. She took the precious stone from her brother's clenched hand, put her hand under his chin and swam with them through this disgusting sludge back to the land.

The people were very upset that Mamlambo had been killed in this way. Now how would they tell stories? How could they create any stories at all now, which had entertained them through the years? They mourned the death of Mamlambo.

Slowly, singing and playing sad music, the Rainbow People made their way back to the compound. Nomvula had an idea. Her idea was to take the skin of Mamlambo once more and to skin the drums. But she also ordered the making of a special drum. It wasn't a drum that had a hollow base like all the others. It was a drum that was made from a seasoned mango tree. It was to be in the shape of a bowl, about the size of the circle made by Nomvula's arms. She had it carved by the best carvers in the Valley of a Thousand Songs. Then she took the story stone and placed it in the middle of the bowl. And she ordered them to stretch the skin of Mamlambo's head over this bowl shape. She took the fangs and pegged the skin in place. The elders named this drum Ngoma—after the festival of the first fruits. They put Ngoma at the center of the drum compound surrounded by all the other drums. This, the most important and most sacred drum, stood there in the center

carved with the symbols of Mamlambo and the Eternal
River of Life. It had to be guarded day and night, in case
anyone tried to steal it.

As for Mkhonto and the Bathakathi, a cure was pre-
scribed for them by the medicine man: their heads were
shaved and smeared with fresh elephant droppings (which
they themselves had to hunt for among the anthills), and
they were made to scrub the compound floor on their
knees. Once they had finished, they were given lessons in
the ancient art of carving passages from the great stories
onto the other drums, and painting the walls of the compound in the colors of the rainbow.

Every morning now Nomvula takes her drum to the
banks of the poisoned river and plays it, dedicating her
music to the memory of Mamlambo. And Mantaba's rainbow smiles at her as every day the Rainbow River flows a
little cleaner, a little clearer. And of course now she can
remember stories, because the story stone is in the sacred
drum. It is now this drum called Ngoma that brings the rain
and gives the Rainbow People their stories.

LIBBY LARSEN

from *Symphony: Water Music*

Measures 68–74, Mvt. II, from *Symphony: Water Music.* Composed by Libby Larsen. © 1984 by E. C. Schirmer Music Company, a division of ECS Publishing, Boston, MA. Used by Permission.

About the Contributors

Samuel Adler is a composer of more than 400 published works, as well as a conductor, author and teacher. He was born in 1928 in Mannheim, Germany, and immigrated to the United States in 1939. His major influences include composers Aaron Copland, Walter Piston, Paul Hindemith and Randall Thompson and conductor Serge Koussevitzky. He has composed operas, symphonies, oratorios and other orchestral chamber works and songs. Adler has served as conductor for many major symphony orchestras in the United States and abroad. He has taught at The Juilliard School since 1997. Former posts include the University of North Texas, the Eastman School of Music and major music festivals in the United States and abroad. In May 2001 he was inducted into the American Academy of Arts and Letters. He is the author of three books: *The Study of Orchestration, Sight Singing* and *Choral Conducting*.

Vladimir Ashkenazy, world-renowned classical pianist and conductor, was born in Gorky in 1937. He studied in Moscow and, after winning second prize at the 1955 Chopin Competition and first prizes at the 1956 Queen Elizabeth and 1962 Tchaikovsky Competitions, went on to perform throughout the world. Since the 1970s, he has become increasingly active as a conductor, holding positions with the Philharmonia (Principal Guest Conductor and now Conductor Laureate), the Royal Philharmonic (Music Director), the Cleveland (Principal Guest Conductor) and Deutsches Symphonie-Orchester Berlin (Chief Conductor and Music Director), and most recently with the Czech Philharmonic (Chief Conductor until June 2003) and the European Union Youth Orchestra (Music Director). Beginning with the 2004–05 season, he will serve as Music Director of NHK Symphony. Ashkenazy has recorded almost all of the major works of the piano repertoire, including the 1999 Grammy Award winning complete Shostakovich 24 Preludes and Fugues. When not at home in Switzerland he appears at the podium and the piano throughout the world.

Emanuel Ax, classical pianist, is renowned not only for his poetic temperament and unsurpassed virtuosity, but also for the exceptional breadth of his performing activity. Ax captured public attention in 1974 when, at age 25, he won the first Arthur Rubinstein International Piano Competition in Tel Aviv. In 1979 he took the coveted Avery Fisher Prize. His distinguished career has included appearances with major symphony orchestras worldwide, recitals in the most celebrated concert halls, a variety of chamber music collaborations, the commissioning and performance of new music and additions to his acclaimed discography on Sony Classical. Ax regularly collaborates with such artists as Yo-Yo Ma, Peter Serkin and Jaime Laredo. As a duo with Ma he has won three Grammy Awards for the sonatas of Beethoven and Brahms. Born in Lvov, Poland, Ax resides in New York with his wife, pianist Yoko Nozaki, and their two children.

Patricia Barber, jazz composer, arranger, pianist and vocalist, is highly acclaimed by critics for her own compositions as well as sophisticated renditions of standards. She was born near Chicago to saxophone-playing father Floyd "Shim" Barber, who had performed with Glenn Miller. After graduation from the University of Iowa in 1979, she launched her career in 1984 in Chicago, first at the Gold Star Sardine Bar and then at the Green Mill. Her 1994 CD *café blue* won *Stereophile*'s designation "Record to Die For" and earned Barber the top spot in the "Female Vocalist" category in the *Down Beat* International Critics Poll, an honor she consistently claims. Barber has frequented *Billboard*'s Top 10 jazz chart with subsequent albums *modern cool, Nightclub* and *Companion*. Her latest release, *Verse,* is a recording of original material. Barber, whose "throaty, come-hither vocals and coolly incisive piano are displayed to devastating effect" (*Time*), tours extensively around the United States and Europe.

Keola Beamer is a multi-award-winning Hawaiian slack key guitarist. He was born into a family whose ancestry dates back to 15th-century Hawaiian royalty and has produced many generations of performing artists. The family revered musical expression as a way to preserve information and communicate with the gods. Beamer began playing guitar, piano and bamboo nose flute as a young boy. In the 1970s, his contributions to slack key guitar (*ki ho'alu*) led to a widespread revival of the tradition. His recordings have reached *Billboard*'s Top 15 world music chart, and he holds the record for the best-selling album in Hawaiian history. Beamer is especially noted for his ability to recontextualize ancient Hawaiian songs into contemporary settings, creating his own unique style. Each performance is a three-dimensional experience, combining the elements of *mele* (song), *hula* (dance) and *oli* (chant) with native instruments and folklore. He continues to teach extensively and perform on the international scene.

Jane Ira Bloom, soprano saxophonist and composer, has been developing her unique style for over 20 years. She is a pioneer in the use of live electronics and movement in jazz. Her continuing commitment to pushing the envelope in her music has led to collaborations with other outstanding jazz and classical artists. Winner of the 2001 Jazz Journalists Award for soprano sax of the year, the *Down Beat* International Critics Poll for soprano saxophone, the IAJE (International Association of Jazz Educators) Charlie Parker Fellowship for jazz innovation and the International Women in Jazz, Jazz Masters Award, Bloom is the first musician ever commissioned by the NASA Art Program. She is a recent winner of the artists' fellowship for jazz composition offered by the Chamber Music America/Doris Duke Project for *Chasing Paint,* a series of compositions for her quartet inspired by the paintings of Jackson Pollock. Bloom is on the faculty of the New School University in New York City.

Gordon Bok, poet, composer, storyteller and folksinger, grew up around the boatyards of Camden, Maine. As he learned to sail, he sang: songs and ballads of the sea and the schooners, the fishes and the fishermen. Later he sang of mythical sea folk, seals and selkies he found in dreams and legends. Bok has been a leader in preserving, collecting, creating and sharing a wide variety of rich and intensely beautiful songs of both land and sea. His mastery of six- and 12-string guitars, his "Bok whistle" and his trademark, the 'cellamba, have added to his well-developed vocal expression. He performs as a solo artist and in concert with others in the United States, Canada, Great Britain and Scandinavia. His music is found in films and folk-music anthologies. Bok, who has recorded more than a score of solo and group albums, still resides in Camden.

Dave Brubeck, jazz pianist and composer, received a Jazz Masters Award in 1999 from the National Endowment for the Arts, and his musical contributions as both pianist and composer over the past half-century have been honored with a Grammy Lifetime Achievement Award. Born in Concord, California, in 1920, young Brubeck worked with his father on a 45,000-acre cattle ranch. Brubeck was drawn to music, especially jazz, and began playing in local bands in his teens. After graduation from college, he enlisted in the Army and served under General Patton in Europe. After World War II, he studied with French composer Darius Milhaud. He has recorded with many of the legendary jazz greats and has performed around the world with the Dave Brubeck Quartet. He has received honorary doctorates from six American and two European universities. He is the recipient of the National Medal of Arts, awarded by President Bill Clinton.

Sarah Chang, violinist, is recognized the world over as one of classical music's most gifted artists. Appearing in the music capitals of Asia, Europe and the Americas, she has collaborated with nearly every major orchestra and with such artists as Isaac Stern, Pinchas Zukerman and Martha Argerich. Notable recital engagements include her Carnegie Hall debut and performances at the Kennedy Center in Washington, D.C., the Barbican Centre in London and the Concertgebouw in Amsterdam. Among her latest releases for EMI is an album of popular shorter works for violin and orchestra, with Plácido Domingo conducting the Berlin Philharmonic. Born in Philadelphia to Korean parents, Chang began to study the violin at age four and soon performed with several area orchestras. Auditions at age eight led to immediate engagements with the New York Philharmonic and the Philadelphia Orchestra. Her remarkable accomplishments to date were recognized in 1999 when she received the prestigious Avery Fisher Prize.

Bruce Cockburn, singer/songwriter, was born in Ottawa in 1945, and quickly was drawn to music as a career. He attended the Berklee College of Music in the early 1960s, but migrated back to Ottawa in 1965 to play rock 'n' roll. He eventually found his voice as a songwriter, drawing on instinctive spirituality, a keen eye for detail and a wry sense of humor. By then he had also developed a highly personal finger-picking guitar style that merged Mississippi John Hurt blues with modal jazz harmony, as well as melodic lyricism and cycling rhythms that suggested an ear for Indian, Asian and African music. "The whole point of writing songs is to share experiences with people," Cockburn says, looking back on a career that includes 26 albums; numerous international awards, including the Canadian Music Hall of Fame and the Tenco Award for Lifetime Achievement in Italy; 20 gold and platinum records in Canada; and countless concert performances. His collected work is a journey both moody and revelatory, with unflinching observations of human cruelty, greed, courage and survival through faith.

Basia Danilow, violinist, enjoys performing internationally. She has appeared in recital at Lincoln Center, Carnegie Hall, Merkin Hall and the Kosciuszko Foundation as well as abroad in Yugoslavia and Russia. Danilow is a winner of the Artists International Competition and a member of the Ariadne Trio and performs and records as guest artist with the Perspectives Ensemble and the Harmonie Ensemble of New York. She is concertmaster of the New Philharmonic of New Jersey and the Princeton Symphony, and performs in the United States and abroad with L'Opera Français, the Orchestra of St. Luke's and the Metropolitan Opera Orchestra. Her numerous festival appearances include Caramoor, Amadeus, Summit, the Central Vermont and Windham Chamber Music Festivals and the International Summer Institute at the Moscow Conservatory. Danilow has recorded for Sony, Atlantic and RCA Victor Red Seal.

Bob Dorough, jazz pianist and singer, is a master of vocalese, the art of writing and singing lyrics to instrumental jazz solos. He was born in 1923 in rural Arkansas. He began playing jazz piano in the 1940s and started singing in the early 1950s. On his 1956 album he introduced his lyrics to Charlie Parker's "Yardbird Suite," and in 1962 he was commissioned by Miles Davis to write "Blue Xmas." He has recorded with such artists as Phil Woods and Billy Hart. His song "Comin' Home Baby" became a hit for Mel Tormé, and his song "Devil May Care" was most recently covered by Diana Krall. Younger generations will recognize Dorough as the composer, arranger, instrumentalist, singer and conductor for the music of *Schoolhouse Rock*. *Down Beat* magazine has praised Dorough for having "the special gift of bringing joy to even the jaded. Perhaps especially to the jaded."

Renée Fleming is a Grammy-Award-winning soprano recognized worldwide for her compelling artistry, beautiful sound and interpretive talents. She continues to be heralded throughout the world by the public and press alike as one of the truly magnificent voices of our time. A champion of new music as well as the standard repertoire, she has performed many world premieres and her voice has resounded throughout distinguished venues worldwide and on numerous recordings. She has performed with most of today's preeminent orchestras and conductors. Fleming's achievements within the classical music industry have been recognized with numerous honors. Outside the classical music world, Fleming's artistry has been acknowledged also; she has appeared in advertising campaigns and served as an inspiration for award-winning novelists. Fleming studied at The Juilliard School and holds degrees from the State University of New York at Potsdam and the Eastman School of Music.

Pamela Frank, classical violinist, has established an outstanding international reputation across an unusually varied range of performing activity. In addition to her extensive schedule of engagements with prestigious orchestras throughout the world and recitals on the leading concert stages, she is regularly sought after as a chamber music partner by today's most distinguished soloists and ensembles. Besides maintaining her partnership with her father, pianist Claude Frank, she works regularly with pianist Peter Serkin and violinist and husband Alexander Simionescu. While committed to the standard classical repertoire, Frank also has an affinity for contemporary music, often including works by today's composers in her programs. For Sony Classical, she has recorded the Chopin Piano Trio and Schubert's "Trout Quintet" with Emanuel Ax and Yo-Yo Ma, and her playing is featured in the film *Immortal Beloved*. The breadth of her accomplishment and her high level of musicianship were recognized in 1999 with the Avery Fisher Prize.

Alan Gampel made his professional classical piano debut at the Hollywood Bowl at age seven. He won the Special Award in the Arts from UNICEF at eight and the Presidential Scholar's Award at the White House in the mid-1980s at 16. After graduating from Stanford University at age 19, he moved to Paris. In 1995, Gampel received the coveted Chopin Prize at the Arthur Rubinstein International Piano Competition. He won a top prize at the Naumburg International Piano Competition, and he was unanimously awarded the Special Mozart Bicentenary Prize. He has appeared at the Kennedy Center, the Louvre, Wigmore Hall and the Mostly Mozart and Ravinia Festivals, and with the Chicago Symphony, the Royal Philharmonic and the Orchestre de Paris, among others. He created the Joy2Learn Foundation, dedicated to bringing the fine arts into classrooms via the Internet.

Evelyn Glennie is the world's first full-time solo percussionist. Born in Aberdeen, Scotland, Glennie graduated from the Royal Academy of Music in London. She has produced 17 albums since launching her professional career, with her first album winning a Grammy. Subsequently, two more albums have been nominated, one, a collaboration with banjo player Béla Fleck, winning again in 2002. At the age of 21, Glennie was awarded the OBE (Officer of the British Empire) for her services to music. Glennie has played with many of the world's top orchestras and has performed in over 40 countries. She typically gives around 110 concerts a year and has commissioned over 100 works for solo percussion from many of the world's most talented composers. Outside of performing, Glennie is committed to improving music education in schools. She has also developed a range of cymbals mallets and even has a tartan named after her: the Rhythms of Evelyn Glennie.

Mike Gordon came to international prominence as a bassist, vocalist and composer with the band Phish. Boston-born Gordon has helped lead the rock band from its beginnings at the University of Vermont to its place as one of the top concert draws in the United States. Through its more than a dozen best-selling albums and thousands of concerts, Phish has earned a reputation for musical imagination, instrumental chops and improvisational experimentation through two decades of intense public popularity. Phish also created the Waterwheel Foundation, which directly helps to protect water. On his own, Gordon released a duet album with legendary guitarist Leo Kottke; an award-winning narrative feature film *Outside Out,* in 2000; and *Rising Low* in 2002, a documentary feature film about 25 famous rock bass players who contributed to an album honoring bassist Allen Woody. Gordon's book of ultra-short stories, *Mike's Corner,* was published in 1997 by Little, Brown's Bulfinch Press.

David Harrington is the founder, artistic director and first violinist of the Kronos Quartet. Synonymous with musical innovation, the Kronos Quartet is known for its unique artistic vision and fearless dedication to experimentation. Since its 1973 inception, Kronos has commissioned many new works, and more than 450 pieces have been written and arranged for the group over the past 30 years. The quartet's extensive repertoire ranges from Alban Berg, Hildegard von Bingen, Charles Mingus and Astor Piazzolla, to Morton Feldman and Carlos Paredes. Kronos tours extensively, with more than 100 concerts a year in concert halls and clubs and at jazz festivals throughout the United States, Canada, Europe, Asia, South America, Mexico, Russia, Australia and elsewhere. The group has won numerous international awards, including three Edison Prizes and eight ASCAP/Chamber Music America Awards for Adventurous Programming and has recorded more than 30 albums.

Mickey Hart, rhythmist, is a social activist and consummate musician best known for his nearly three decades as an integral part of the rock band the Grateful Dead. He was half of the percussion tandem known as the Rhythm Devils, a duo whose extended polyrhythmic excursions were highlights of Grateful Dead shows. In those shows, Hart introduced audiences to many percussive instruments from around the world. One of the most successful of his dozens of albums, Hart's *Planet Drum,* was the number-one hit on the *Billboard* world music chart for 26 weeks and received the Grammy for Best World Music Album. Hart has composed scores, soundtracks and themes for movies, television and home video, including *Apocalypse Now, The Twilight Zone, Hearts of Darkness* and others. He is very involved with Save Our Sounds and is currently completing a book called *Songcatching* to be published by National Geographic. He continues to investigate the connection between healing and rhythm.

Tilla Henkins and her husband, Francois, are products of the first orchestral training program in South Africa, which started 42 years ago. They are still passionate about developing music and helping it survive challenging times by intensively training musicians and organizing and playing in chamber music and symphony orchestra concerts. Henkins has a class of 25 cellists at the Free State Musicon in Bloemfontein. She uses many of her own arrangements in the string quartets, the cello ensembles and the string orchestra she is training. She wants to make lots of indigenous music available to music-teaching programs in South Africa soon. She also challenges herself to develop students' musical and technical skills through encouragement, experimentation and edification, her personal motto being that "no form of criticism is constructive." The Henkins family lives on a small farm, where they have raised four excellent young musicians amongst a number of sheep, dogs and horses.

Sharon Isbin is a Grammy Award–winning classical guitarist acclaimed for her extraordinary lyricism, technique and versatility. "The preeminent guitarist of our time" (*Boston Magazine*), Isbin has won the Toronto, Munich and Madrid International Competitions and *Guitar Player*'s "Best Classical Guitarist" and "Album of the Year" awards. She is the first classical guitarist to win a Grammy in 28 years, awarded for her solo CD *Dreams of a World* (Teldec Classics). She gives sold-out performances in the world's greatest halls, including New York's Avery Fisher and Carnegie Hall, Boston's

Symphony Hall, Washington D.C.'s Kennedy Center, Madrid's Teatro Real, London's Barbican Centre and Amsterdam's Concertgebouw. She has toured Europe annually since she was 17 and has soloed with more than 140 orchestras worldwide. Her many albums, from Baroque, Spanish/Latin and 20th-century repertoire to crossover and jazz-fusion, are frequent bestsellers. She is a book author and the director of the guitar departments at the Aspen Music Festival and Juilliard. Website: www.sharonisbin.com

Paavo Järvi is an internationally sought-after conductor who champions the classical music of Scandinavia and his native Estonia. As former principal guest conductor of the Royal Stockholm Philharmonic and the City of Birmingham Symphony, he recorded extensively for EMI/Virgin Classics, including the music of Sibelius, Pärt, Tüür, Stenhammar and Leonard Bernstein. In autumn 2001 Järvi assumed the post of music director of the Cincinnati Symphony Orchestra, with which he has released a Berlioz disc. As guest conductor Järvi regularly leads the world's top orchestras and often returns to Estonia, where he conducts the Estonian National Symphony in concert and recording. He also endeavors to work with young musicians, and has led the Verbier and European Union Youth Orchestras on European festival tours. He studied percussion and conducting at the Tallinn School of Music and moved to the United States in 1980 to study at Curtis Institute of Music and with Leonard Bernstein at the Los Angeles Philharmonic Institute.

Wesley Jefferson, the "Mississippi Junebug," is a blues singer, songwriter and teacher who calls Clarksdale, Mississippi, home. "Junebug," as he is affectionately known in Clarksdale, was born March 23, 1944, in nearby Roundaway, Mississippi. Like many black families of the era, his family moved from plantation to plantation, trying to make a living sharecropping. At a young age, Jefferson left school to pick cotton in an effort to help support his family after his father left. During the 1970s, 1980s and 1990s, Jefferson was central to Clarksdale's blues scene. For years he held court at Smitty's Red Top Lounge and later at Margaret's Blue Diamond Lounge. Today, he still plays the blues and also helps manage nine talented children who make up the blues band Blues Prodigy. His warm bass playing and vocals are featured on his album *The Wesley Jefferson Band Sings the Blues: Live from the Do Drop Inn.*

Tunde Jegede is a composer, cellist and performer of the 21-stringed Malian harp-lute (kora) from West Africa. He uniquely bridges African and European classical musical traditions, having a foundation and being rooted in both of them. Born in 1972, Jegede was introduced to the kora by virtuoso Senegalese musician Bouly Cissokho. At 10, Jegede was invited to study within the ancient Griot tradition with the Master Amadu Bansang Jobarteh. He later studied the cello and Western classical music with Russian cellist Alfia Bekova. As pioneers of African classical music, Jegede and his ensemble have given numerous recitals and performances internationally in concert, at festivals and on radio and television. He has composed for leading orchestras and performers and contributed to many albums, including *War of a Beach Goddess, River of Sound, Flute for Thought* and *Living Magic.* His own compositions include *Lamentation* and *Malian Royal Court Music.*

Jaime Laredo, classical violinist and conductor, was born in Bolivia. Since his stunning orchestral debut at the age of 11 with the San Francisco Symphony, he has won the admiration and respect of audiences, critics and fellow musicians with his passionate and polished performances. Laredo, who debuted at Carnegie Hall over 40 years ago, has excelled in the multiple roles of soloist, conductor, recitalist and chamber musician. He juggles a complex musical schedule as artistic director of New York's renowned Chamber Music at the Y series, as music director of the Vermont Symphony Orchestra, and as a member of the celebrated Kalichstein-Laredo-Robinson Trio with multiple solo and ensemble performances. He has recorded more than 100 discs and received seven Grammy nominations, winning for a disc of Brahms Piano Quartets on which he performed with longtime friends Isaac Stern, Yo-Yo Ma, and Emanuel Ax. He resides with wife, cellist Sharon Robinson, in Vermont.

Libby Larsen, one of the most celebrated composers working today, has created an immense catalogue of works that spans virtually every genre. Born in 1950, and a co-founder of the American Composer's Forum in 1973, Larsen, through her music and ideas, has refreshed the concert music tradition and the composer's role in it. Her awards and accolades are numerous, including a 1994 Grammy for the recording of *Sonnets from the Portuguese.* A discography of over 48 works reflects her many friendships and collaborations with world-renowned artists. *Gramophone* called her *Symphony: Water Music* "the finest water music since Respighi's *Fountains.*" She has served as composer in residence for many leading symphonies, and her writings and speeches on music can be found in numerous textbooks. Two of Larsen's recently released compositions are *Love After 1950,* performed and recorded by mezzo-soprano Susanne

Mentzer, and the Colorado Symphony's world premiere of her fifth symphony, *Solo Symphony.*

Kenny Loggins is one of the most successful songwriters and performers in the world. With a career spanning more than thirty years, he has logged in twelve platinum albums and fourteen gold albums in the United States alone, and has sold more than twenty million albums worldwide. His title track for the film *Footloose* earned him an Academy Award Nomination, and along the way he picked up two Grammy Awards. Kenny's passion for environmental issues resulted in the television special and accompanying live album *Outside: From The Redwoods,* and his environmental TV special "This Island Earth" was awarded two Emmy Awards. Kenny is a visible role model of how to seamlessly combine career, fatherhood, conscious loving, and personal growth in your life. Kenny is the father of five wonderful children.

Iain Mac Harg is an accomplished bagpiper whose introduction to the instrument came from his father, a premier bagpipe builder. Mac Harg's instructors in the pipes include P. M. George, F. Ritchie, Donald Lindsay, Bruce Gandy, Scott Mac Aulay and Andrew Wright. In addition to competing and performing as a solo piper, Mac Harg has been involved in many other aspects of Celtic music, including the founding of two Highland pipe bands. He also plays with several folk groups and has been a familiar face at the highland games since very early in his life. After college graduation in 1997, Mac Harg began to develop the Vermont Institute of Piping, modeling it after the College of Piping in Prince Edward Island. Mac Harg is Vermont's only full-time piping instructor. His solo recordings include *Rooted in Tradition* and *Celtic Christmas.* He has also published a collection of original tunes for the bagpipe.

Taj Mahal first made his mark on the musical scene when he and Ry Cooder co-founded the group the Rising Sons, which quickly earned Mahal a recording contract. He has since recorded 39 albums of American originals and earned six Grammy nominations and one Grammy Award, for his 1997 *Señor Blues.* His career features side trips into Chicago blues, Memphis soul and West African, Caribbean, Cajun and Latin sounds. He has allied himself with a host of musicians on an array of diverse musical projects, particularly 1999's critically acclaimed *Kulanjan* collaboration with West African kora master Toumani Diabate. Mahal plays more than 20 instruments, including the National Steel and Dobro guitars. He has recorded a handful of film soundtracks and has made multiple big-screen appearances in several major motion pictures. Mahal has embraced multiple musical traditions in his nearly 40-year career but always returns to the blues.

Carol Maillard is an original member of Sweet Honey In The Rock. As a vocalist, she also can be heard with Horace Silver on *Music of the Sphere,* and on *Betty Buckley at Carnegie Hall* and *Sounds of Light.* Her arrangement and lead performance of the spiritual "Motherless Child" can be found on the soundtrack for *The Visit.* Maillard is an accomplished actress in theater, commercials and cabaret, performing in the premiere season of American Playhouse on PBS, the D.C. Black Repertory Company, the New York Shakespeare Festival, the Negro Ensemble Company, the Actors Studio and the Amas and San Diego Repertory Theaters. She has appeared in feature films *Beloved* and *Thirty to Life* and on the television shows *for colored girls who have considered suicide* and *Hallelujah!* Her on- and off-Broadway credits include performances in *Eubie!, Don't Get God Started, Comin' Uptown, It's So Nice to Be Civilized, Beehive* and *Forever My Darling.*

Mischa Maisky has the distinction of being the only cellist in the world to have studied with both Mstislav Rostropovich and Gregor Piatigorsky. Russian-born and -educated, Maisky has performed in concert halls and music centers around the world. His musical partnerships have included such artists as Martha Argerich, Radu Lupu, Zubin Mehta, Gideon Kremer, Yuri Bashmet, Leonard Bernstein, Vladimir Ashkenazy, Daniel Barenboim and many others. Maisky's highly acclaimed recordings have been bestsellers worldwide and have earned him the Record Academy Prize in Tokyo three times and the coveted Grand Prix du Disque in Paris. Among his many distinctive recordings are the Shostakovich concertos with the London Symphony and Michael Tilson Thomas, the Vivaldi and Boccherini concertos with the Orpheus Chamber Orchestra and the Bach Suites for Solo Cello. His release of the works of Saint-Saëns received France's highest music award, the Diapason d'Or. He resides in Belgium with his family.

Mary Elizabeth, composer, studied briefly with Easley Blackwood at the University of Chicago. Otherwise she is self-taught, but has benefited from opportunities and support provided by Jerome Monachino, Xan Johnson, D. Thomas Toner, Troy Peters and others. She first focused on *a cappella* and instrumentally accompanied songs, with performances by the St. Michael's College Liturgical Choir, the University Choral Union, the Essex Children's Choir and a premiere of settings of sixteenth-century Spanish poetry sung by the composer accompanied by harpist Heidi Soons at the Center for the Arts at Middlebury College. Recently, she has composed chiefly for theater, including the Zona Gale Center's Youth Theatre Ho-Chunk Project, *Keepers of Harmony,* which was showcased at the American Alliance for Theatre and Education conference in 2002, and for the theater-production *The Haunted Forest 2002.* Currently, she is composing a ballet, and the Vermont Contemporary Music Ensemble has plans to perform selections from her settings of John Engels' poetry.

Bobby McFerrin was born to opera-singer parents in New York in 1950. A 10-time Grammy winner, McFerrin is one of the world's best-known innovators and improvisers, a world-renowned classical conductor, the creator of one of the most popular songs of the late 20th century ("Don't Worry, Be Happy") and a passionate spokesman for music education. His recordings have sold over 20 million copies, and his collaborations with Yo-Yo Ma, Chick Corea, the Vienna Philharmonic and Herbie Hancock have established him as an ambassador of both the classical and jazz worlds. With a four-octave range and a vast array of vocal techniques, McFerrin is a vocal explorer who has combined jazz, folk and a multitude of world music influences—choral, a cappella and classical music—with his own ingredients. Yet his reach extends well beyond musical circles; he has worked with actor Jack Nicholson, comedians Robin Williams and Billy Crystal, the Muppets and audiences around the world.

Brad Mehldau is a groundbreaking jazz pianist who began experimenting with the piano at age four, long before he was exposed to jazz, and began taking lessons when he was six, continuing only until he was 14. He turned to jazz piano when he moved in 1988 to New York City, where he recorded a series of albums as a sideman with a variety of leading musicians. Mehldau's first major international exposure came as a member of the Joshua Redman Quartet, with which he recorded *MoodSwing.* While in New York, Mehldau met and began to work with his future trio members, Larry Grenadier and Jorge Rossy. The Brad Mehldau Trio now performs around the world, and has recorded many highly acclaimed albums together. In response to his debut album, *New York, Newsday* hailed Mehldau as "among the most compelling, eccentric and daring young pianists in jazz." *Art of the Trio 1* garnered Mehldau a 1998 Grammy nomination for Best Jazz Instrumental. His music is featured in the film *Eyes Wide Shut.*

Susanne Mentzer, mezzo-soprano, is the rare singer with equal vocal and acting gifts. She is a familiar face at the world's most prestigious opera houses, including the Metropolitan Opera, La Scala, the Vienna State Opera and the Opéra de Paris, and at such festivals as Salzburg, Tanglewood and Montreux. She excels in challenging title roles, ranging from the classic femme fatale to ingénue and "trouser," and her voice captures Mozart, Debussy and Richard Strauss as fluidly as the *bel canto* style of Rossini, Bellini and Donizetti. Born in Philadelphia and raised in Maryland and New Mexico, Mentzer studied voice at Juilliard and the Houston Opera Studio and privately with Norma Newton. She is on the faculty at both DePaul University School of Music and the Aspen Music Festival. Since 1992, Mentzer has organized the annual Jubilate benefit to support Chicago's Bonaventure House, a residence for people living with AIDS. She is the proud mother of a teenage son.

Jim Messina is a guitarist, writer, singer and producer. A member of such seminal groups as Buffalo Springfield and Poco, Messina embarked on his rock career as a producer for Lee Michaels, Joni Mitchell and the Doors. In the 1970s Messina and Kenny Loggins joined forces and recorded nine albums over seven years, amassing sales of over 14 million units. Since then, Messina has worked as a solo artist and released the retrospective album *Watching the River Run* in 1996. Today he continues to write, sing and produce from his studio in the Santa Ynez Valley and conduct the Songwriters' Performance Workshop across the United States.

Midori, heralded as one of the world's foremost violinists, balances her schedule between solo recitals, concerto appearances with orchestras throughout the world and a dedication to outreach programs. She founded her own organization, Midori & Friends, in 1992 to bring diverse styles of music and music instruction to children in public elementary schools at no cost to students. An exclusive Sony Classical recording artist, she has recorded the concertos of Bartók, Dvořák, Shostakovich, Tchaikovsky and Sibelius, as well as several recital albums with pianist Robert McDonald. Midori was born in Osaka, Japan, in 1971 and began violin studies with her mother, Setsu Goto, at age four. She plays the 1734 Guarnerius del Gesu "ex-Hubermann," on lifetime loan to her from the Hayashibara Foundation. She lives in New York City.

Yousuke Miura, singer, was born in 1950 in Kumamoto prefecture in southern Japan. He formed the male duo Again a decade ago with friend Sadao Kojima. Performing professionally since 1996, the two primarily sing Japanese folk songs about nature, the seasons and people's lives, accompanying themselves on guitar. In addition to performing in concert halls in Japan and the United States, the two artists sing for elderly people, the disadvantaged and school-children. Many of their songs are drawn from old children's music books that have disappeared from use, or from compositions from the post–World War II era in Japan. After they began to sing for the elderly, Miura and Kojima discovered that such singing can actually help cure people. The duo also performs songs unfamiliar to the youth in Japan in the hopes that the songs will survive for generations to come.

Keb' Mo', a Los Angeles–based musician, communicates with absolute authenticity, heart to heart. Supported by a cast of other brilliant musicians, Mo's riveting guitar work, particularly on slide guitar, is a distinguishing feature of his music. He is a triple winner of the 1999 W. C. Handy Blues Awards and a winner of two Grammys, for his albums *Slow Down* and *Just Like You.* Born Kevin Moore, his performance name grew from a street-talk version of his original name. As a teenager, he also blew trumpet and French horn. In his first group, a calypso band, he also played upright bass and steel drums. As he turned increasingly to the blues, he found his niche. Through playing with musicians as varied as Papa John Creach, the Rose Brothers, the Whodunit Band, Big Joe Turner, Pee Wee Crayton, Eugene Powell and others, he developed his distinctive style as a blues guitarist.

Randy Newman, born in Los Angeles in 1943, grew up watching his Uncles Lionel, Emil and Alfred compose and record music for film on the Twentieth Century Fox sound stage. By the time he was 17 he was writing pop songs for the music publishing house, Metric Music, and in 1968 released his first album "Randy Newman," for Reprise Records. His songs have been recorded by Judy Collins, Three Dog Night, Joe Cocker, Dusty Springfield and many others. As a film composer, he has received 16 Academy Award nominations for *Ragtime, The Natural, Avalon* and *Toy Story,* among others. In 2002, he won an Oscar for the song, "If I Didn't Have You," from the film *Monsters, Inc.*

Nerissa Nields is a founding member of the Nields, a successful folk-rock band of the 1990s, and a current member of the duo Nerissa and Katryna Nields. Launching its career in New England coffee-houses in 1991, by 1995 the Nields, a five-piece folk-rock band, had begun touring nationally. Between 1994 and 2000, the Nields released six full-length CDs, and performed over 200 concerts a year throughout North America. With triple-A and college radio embracing the band, the Nields sold close to 100,000 albums. In

the summer of 1998, Nerissa and Katryna Nields joined the Lilith Fair. Within a year, the duo was opening for national acts and playing the Newport Folk Festival. Nerissa Nields has written her first young-adult novel, *This Town Is Wrong.* It is scheduled for release in fall 2003.

Erik Nielsen has been composing for 30 years. His catalogue includes works for chorus, orchestra and solo instruments, chamber music and electronic music. His works have been performed in Europe and Australia as well as in many locations across the United States. Numerous ensembles have performed his work, including the Emerson and Ying String Quartets, the Bread and Puppet Theater, the Vermont Contemporary Music Ensemble, the Vermont Symphony and Village Harmony. He has won awards from ASCAP, the Vermont Music Teachers Association and the Vermont Arts Council. His 1995 piano quintet was performed at Carnegie Hall by the Manchester Chamber Players, and in 2000 his opera, *A Fleeting Animal: An Opera from Judevine,* premiered to great acclaim. He was recently commissioned by the National Symphony Orchestra to write a chamber work to be premiered at the Kennedy Center in Washington, D.C.

Mark O'Connor, Grammy-winning violinist and composer, is widely acknowledged as one of the finest musicians of his generation. He tours the country appearing with major symphony orchestras and in solo recital. He has performed at the White House for U.S. presidents, for the centennial Olympics celebrations and with other leading musicians including Yo-Yo Ma, Pinchas Zukerman, Wynton Marsalis and the late jazz great Stephane Grappelli. O'Connor has received commissions from the prestigious Meet The Composer program and from the Library of Congress, for which he composed a sonata in celebration of the institution's centennial celebration. The Mark O'Connor Fiddle Camps and string conferences, held twice a year, attract faculty and students from all over the world. He has been featured on *CBS News Sunday Morning* and NPR's *Performance Today* as well as on numerous radio and television broadcast specials.

Garrick Ohlsson, since winning the 1970 Chopin International Piano Competition in Warsaw, has established himself as a classical pianist of extraordinary interpretive power and prodigious technical facility. Although regarded as one of the world's leading exponents of the music of Chopin (Ohlsson has recorded the complete solo works, in 13 volumes), he commands an enormous repertoire that encompasses virtually the entire piano literature. He is likewise acclaimed for his performances of Mozart, Beethoven, Schubert and the Romantic repertoire. Ohlsson performs regularly with the world's great orchestras and maintains an unusually active recital schedule. He was born in White Plains, New York and entered The Juilliard School at age 13. He won first prize at the 1966 Busoni Competition in Italy and at the 1968 Montreal Piano Competition, but it was his Chopin triumph that made him a favorite in Poland, where he has toured in performance numerous times.

Christopher Parkening, one of the world's preeminent virtuosos of the classical guitar, is an heir to the legacy of the great Spanish guitarist Andrés Segovia. Parkening's extensive concerts, recitals and recordings have captivated listeners around the world for more than a quarter-century. He has performed with the finest orchestras in the United States as well as at the White House, at Carnegie Hall's 100th-anniversary celebration and twice on the internationally televised Grammy Awards. Parkening has amassed an extensive discography and has received two Grammy nominations for Best Classical Recording and the American Academy of Achievement Award. He has authored *The Christopher Parkening Guitar Method,* volumes 1 and 2, and a collection of guitar transcriptions and arrangements. Each summer, Parkening teaches music classes at Montana State University. He is a world fly-fishing and casting champion and resides with his wife, Theresa, in Southern California.

William Parker, bassist, composer and improviser, has been active on the creative-music scene both nationally and internationally since 1972. As a bassist he has played with Don Cherry, Cecil Taylor, Bill Dixon, Milford Graves, Matthew Shipp and Davis S. Ware, as well as with the top musicians in the world of avant garde jazz. Parker is the leader of the Little Huey Creative Music Orchestra. His CDs include *O'Neal's Porch* (Aum Fedelity), *Raining on the Moon* (Thirsty Ear Records), *Bob's Pink Cadillac* (eremite records), *Raincoat in the River* (eremite records), and *Mass for the Healing of the World* (Black Saint Records). He is the author of three books, *The Sound Journal, Music and the Shadow People* and *Document Humanum.*

Hermeto Pascoal, in the rich and diverse universe of music in Brazil, shines like a comet, crossing eras and musical circles, leaving a strong influence on generations of musicians and building a name that represents unbridled creativity and inspiration. Better known as a genial multi-instrumentalist, he is capable of extracting music from the most unexpected objects, while exhibiting his virtuosity on piano, flute, saxophone, strings, percussion and many other conventional and unconventional instruments. His compositions have inspired players like Miles Davis, Gil Evans, Herbie Hancock and many others. Pascoal has led one of the most cohesive ensembles in the world for decades, performing his compositions and featuring himself as soloist. He has written music for a wide variety of ensembles, from solo instruments to symphony orchestras and big bands, and continues to compose daily at his home in Rio de Janeiro.

Marcus Roberts, jazz pianist/composer, was born in Florida in 1963 and was first exposed to music in the local church where his mother was a gospel singer. Losing his sight at age five, Roberts began teaching himself to play piano at age eight, then left home to attend the Florida School for the Blind in St. Augustine at nine. He studied classical piano at Florida State University in Tallahassee. After graduation, he toured with trumpeter Wynton Marsalis for six years. Roberts' recordings have routinely reached number one on the *Billboard* jazz charts as he continues to pursue new ways to blend traditional jazz and classical music. The Marcus Roberts Trio exemplifies his approach to music, but Roberts also performs regularly as a soloist with symphony orchestras around the world. He has been instrumental in the training of many talented young jazz musicians. The award he is most honored to have received is the 1998 Helen Keller Award for Personal Achievement.

Sharon Robinson, cellist, is a Texas native who was born into a musical family and who gave her first concert when she was seven. She has won many awards and honors in the years since for her dynamic performances, including the Avery Fisher Recital Award, the Piatigorsky Memorial Award and a Grammy nomination. Whether Robinson is performing as a recitalist, soloist with orchestra or member of the renowned Kalichstein-Laredo-Robinson Trio, critics and audiences respond to what the *New York Times* calls "an artistic personality that vitalizes everything she plays." She and husband Jaime Laredo serve as co-artistic directors of the Hudson Valley Chamber Music Circle. Her passionate playing has inspired numerous commissions for solo cello, chamber works and concerti by some of the leading composers of the day, including Richard Danielpour, who wrote a double concerto for Robinson and Laredo entitled *In The Arms of the Beloved,* celebrating the duo's 25 years of marriage.

Paula Robison is treasured on several continents for bringing pure intonation, beautifully varied tone and extraordinary technical mastery to classical flute. At age 20 Robison was invited by Leonard Bernstein to be guest soloist with the New York Philharmonic. She was the first American to win first prize at the Geneva International Competition. Robison was a founding member of the Chamber Music Society of Lincoln Center and for 10 years was co-director of chamber music at both the Italian and American Spoleto Festivals. She continues to participate regularly in both these festivals and in the Marlboro Music Festival. Her astonishingly diverse repertoire features her affinity for Brazilian music, heard on two discs, *Brasileirinho* and *Rio Days, Rio Nights.* With a lively interest in expanding the flute repertoire, she has commissioned concertos for flute and orchestra from Leon Kirchner, Toru Takemitsu, Kenneth Frazelle, Oliver Knussen and Robert Beaser. Her books are published by Universal, Schott and European American Music.

John Ruskey began painting as a way of studying the motions and patterns of the muddy currents of the Mississippi River, which express themselves in violent boils, eddies and whirlpools the size of Winnebagos. He has been fascinated with water since infancy. His parents couldn't keep him in the crib because of his attraction to a duck pond across the street in the Denver City Park. Accomplished at singing, songwriting and multiple instruments including the piano and guitar, Ruskey studied blues guitar and keyboards with Johnnie "Mr. Johnnie" Billington after moving to Mississippi. He was the co-founder and director of the Delta Blues Education Program in Clarksdale, Mississippi, a program that provides year-long apprenticeships on the blues for Delta children, and the curator of the Delta Blues Museum from 1992 through 1998. Ruskey writes a monthly column for *Blues Revue Magazine,* conducts guided canoe expeditions on the Lower Mississippi River and builds canoes for the Quapaw Canoe Company.

Marjorie Ryerson is a college professor, journalist, photographer and poet. Her photographs have appeared in such diverse publications as *Vermont Life, The Boston Globe, Yankee, Country Living,* and the photography books, *The Vermont Experience* and *Vermont for Every Season.* As an art photographer, she has had numerous one-woman shows of her work. Her poetry and nonfiction have been published by magazines and newspapers across the Northeast, and she has recently completed the book, *Witnesses at the Gate: Stories of the Intimate Privilege of Accompanying the Dying.*

Ryerson is Professor of Communication at Castleton State College, where she teaches journalism and photography. Each spring she teaches poetry for Middlebury College at the annual New England Young Writers' Conference at Bread Loaf. Ryerson has a Master of Fine Arts degree in poetry from the Writers' Workshop at the University of Iowa.

The publishing of *Water Music* has led to the development of the larger *Water Music* Project (www.WaterMusicProject.com), which Ryerson has established to assist her in raising additional revenue for the protection of water around the globe. Skilled advisors representing the environment, law, finance, literature, education and music are working with Ryerson to create concerts featuring the book's musicians, as well as lectures, the commissioning of new music inspired by water, readings, and art shows. Revenues from the many events will be donated to the same *Water Music* Fund of the United Nations Foundation that the book supports.

Marjorie Ryerson plays classical piano and jazz saxophone and lives in Randolph, Vermont.

Nadja Salerno-Sonnenberg, violinist, was born in Rome and emigrated to the United States at the age of eight to study at the Curtis Institute of Music. One of the world's preeminent instrumentalists, she is among the most dynamic, original and daring artists on the concert stage today, heralded for her unique sound, passionate interpretations and musical depth. She has performed with most of the world's greatest conductors and orchestras and at major international festivals. Salerno-Sonnenberg's versatility and vast range of interpretive skills are further demonstrated in the recording field, where she is considered a groundbreaker. She has been highlighted on a variety of television programs as guest, host and featured actress/musician. She was the subject of a 2000 Academy Award–nominated documentary film on her life, *Speaking in Strings.* Salerno-Sonnenberg's professional career began in 1981 when she won the Naumburg Violin Competition. In 1999 she was honored with the prestigious Avery Fisher Prize.

Peter Sanders, cellist, is a native New Yorker, a member of the New York City Ballet Orchestra (for which he has served as acting principal) and principal cello for the New Philharmonic of New Jersey and also performs with the Riverside Symphony, the Eos Orchestra, the Stamford Symphony and the Perspectives Ensemble, with which he has recorded as a guest artist. He is artistic director of the Central Vermont Chamber Music Festival (www.centralvtchambermusicfest.org), which had its inaugural season in 1993. Sanders is a member of the Ariadne Trio and has participated in many summer festivals including the Colorado Music Festival, Skaneateles Festival, the Crested Butte Chamber Music Festival, the Eastern Music Festival, the Lancaster Festival in Ohio, where he was principal cello from 1992 to 1998, and the Windham Chamber Music Festival. As a studio musician Sanders has recorded for a variety of popular artists including Pat Metheny, Jewel and Kathie Lee Gifford.

Joseph Schwantner, composer, has enjoyed countless commissions from the world's finest orchestras, festivals, institutions and musicians around the globe. Schwantner was composer-in-residence with the Saint Louis Symphony Orchestra as part of the Meet The Composer/Orchestra Residencies Program funded by the Rockefeller Foundation and the National Endowment for the Arts. His music has garnered two Grammy nominations, for *Magabunda/ Four Poems of Agueda Pizarro* and *A Sudden Rainbow,* and a Pulitzer Prize, for the orchestral work *Aftertones of Infinity*. Schwantner was born in Chicago and received his training at the Chicago Conservatory and Northwestern University, completing a doctorate in 1968. He previously served on the faculties of the School of Music at Yale University, the Eastman School of Music and The Juilliard School. He has been the subject of a national public-television documentary entitled *Soundings*. In 2002 he was elected a member of the American Academy of Arts and Letters.

Pete Seeger has been singing old songs and making up new ones since 1939. Once blacklisted from national television for being unafraid to voice his opinions, Seeger wrote, co-wrote or introduced songs that are today part of the American landscape: "We Shall Overcome," "Guantanamera, " "Where Have All the Flowers Gone?" "This Land Is Your Land," "If I Had A Hammer" and "Turn! Turn! Turn!" He received the Kennedy Center Honors in 1994, and in January 1996 he was inducted into the Rock and Roll Hall of Fame. He won a 1996 Grammy for Best Traditional Folk Album for his Living Music recording *Pete*. With wife Toshi he helped found the Clearwater Organization, a group dedicated to cleaning up the Hudson River and to environmental awareness. In April 1999, he traveled to Cuba to accept the Felix Varela Medal, that nation's highest honor, for his work in defense of the environment. He describes himself now as a "songleader."

Russell Sherman, classical pianist, has earned a reputation as a virtuoso and interpreter of remarkable intelligence. He has performed with America's leading orchestras, including Boston, Chicago, Los Angeles, New York, Philadelphia, Pittsburgh and San Francisco,

and in prestigious keyboard series from coast to coast. Abroad, his artistry has been heard in major cities in Europe, South America and Asia. According to the *American Record Guide,* Sherman belongs among the elite of Beethovenians: he has performed and recorded all five Beethoven concerti with the Czech Philharmonic, and is the first American to record both Beethoven's complete piano sonatas and the five piano concertos. The New York native, who graduated from Columbia University at age 19, has been a visiting professor at Harvard University and is a distinguished artist-in-residence at the New England Conservatory. His acclaimed book of essays on piano playing and allied activities, *Piano Pieces,* has been published by Farrar Straus Giroux.

Eugene Skeef, percussionist, composer, poet and conflict-resolution activist, was born in South Africa and moved to London in 1980. He was director of music development for War Child, an international aid network that organizes music workshops for children as part of Yugoslav relief efforts. Under the War Child rubric Skeef was also appointed the first director of music development for the affiliated Pavarotti Music Centre in Bosnia, which launched in 1997. War Child hopes to establish more such international music links, demonstrating the way that music can be used creatively and therapeutically to bring young people together to create a more peaceful future. Skeef's awards include a Certificate of Appreciation from the Bosnian Peace-Building Wilderness Learning Program in Minnesota in 1999, the Queen's Award in 1995, and the ABSA (Association for Business Sponsorship of the Arts) Award in 1995. As a consultant on cultural development, he advises the Contemporary Music Network of the Arts Council of England. He was appointed a fellow of the Royal Society of Arts in 2001.

Richard Stoltzman is a clarinet recitalist, chamber musician, jazz innovator, recording artist and soloist with the world's top orchestras. He has earned an international reputation for expanding audiences for the instrument, including performing the first clarinet recitals in the histories of both the Hollywood Bowl and Carnegie Hall. His jazz recital at the Bayreuth Opera House, the traditional home of Wagner, caused a scandal for the local presenter but brought a cheering, capacity audience that demanded five encores. As a founding member of the chamber music group TASHI, he has premiered works by Toru Takemitsu and Charles Wuorinen with the Boston Symphony and the Cleveland Orchestra. Stoltzman has won two Grammy Awards and the coveted Avery Fisher Prize, the first given to a wind player. Born in Omaha, Nebraska, the son of a jazz-playing railwayman, Stoltzman is married to a concert violinist and has two children, both musicians. He is a Cordon Bleu–trained pastry chef.

Tan Dun, composer, is the winner of the prestigious Grawemeyer Award for the composition of the opera *Marco Polo* and the 2003 Composer of the Year by *Musical America*. He began his musical career with the Peking Opera in China and he attended Beijing's Central Conservatory. Tan Dun holds a doctoral degree from Columbia University, where he studied with Mario Davidovsky and Chou Wen Chung. Tan Dun's music is played throughout the world by the leading orchestras and ensembles of our time. Tan Dun's compositions include the *Orchestral Theatre* series, a four-hour orchestral exploration of multicultural and multimedia programs; *Water Passion After St Matthew* for the Internationale Bachakadamie in Stuttgart, commemorating the 250th anniversary of Bach's death; the *Concerto for Water Percussion and Orchestra* for the New York Philharmonic with Kurt Masur; both Oscar and Grammy-winning film score for Ang Lee's film, *Crouching Tiger, Hidden Dragon*; the opera *Tea*, commissioned by Japan Suntory Hall and directed by Pierre Audi, in a co-production with the Netherlands Opera, and *The Map: Concerto for Cello, Video & Orchestra* composed for the Boston Symphony Orchestra and cellist Yo-Yo Ma. Future commissions include an opera for the Metropolitan Opera (scheduled for 2006) and works for the Los Angeles Philharmonic and the Berlin Philharmonic.

Gregory Turay is a tenor who appears regularly at the Metropolitan Opera, the Lyric Opera of Chicago, the San Francisco Opera and the Opera Theater of St. Louis, and in recitals and concerts throughout the United States and Europe. Several important recent debuts include creating the role of Rodolpho in the world premier of Arthur Miller's *A View from the Bridge,* by Bill Bolcom, and starring roles in Mozart's *Cosi fan tutte* with the Welsh National Opera, *The Magic Flute* with Deutsche Oper Berlin, and Haydn's *Creation* with the Boston Symphony. He won the Metropolitan Opera National Council Auditions in 1995 at the age of 21, the Richard Gaddes Award from the Opera Theatre of St. Louis in 1997 and the 2000 Richard Tucker Award. The University of Kentucky graduate has been hailed by the *Times* of London as "one of the brightest natural talents to have emerged from the U.S. in recent years."

Randy Weston, African rhythms, was born in Brooklyn, New York, in 1926. The young jazz pianist was surrounded by early influences such as Max Roach and Cecil Payne. In the 1950s he established himself as a composer with "Hi-Fly" and began a lifelong collaboration with arranger Melba Liston. An African tour on behalf of the U.S. State Department in the 1960s inspired him to settle in Morocco. Since then, Weston has explored the musical and spiritual connections between jazz, African, classical Chinese and European traditions. He released *Earth Birth* with the Montreal String Orchestra in 1997 and *Spirit!* with his African Rhythms Quintet and the Master Gnawa Musicians of Morocco in 1999. His rich career has been recently honored with the Jazz Masters Award from the National Endowment for the Arts, a Lincoln Center commission, a weeklong residency at Harvard University and the French Order of Arts and Letters.

George Winston, solo pianist, was born in 1949 and spent most of his formative years in Montana. This upbringing, with the distinct changes in the seasons, became his primary inspiration for playing music. His solo piano albums include *Ballads & Blues 1972, Autumn, Winter Into Spring, December, Summer, Forest, Linus & Lucy—The Music of Vince Guaraldi, Plains, Remembrance—A Memorial Benefit,* and *Night Divides the Day—The Music of the Doors*. He continues to tour regularly, as well as recording the masters of the Hawaiian slack key guitar tradition for his label Dancing Cat Records.

Paul Winter is a saxophonist whose award-winning body of work chronicles his wide-ranging experiences in the musical traditions and natural environments of the Earth. His "Earth Music" embraces the traditions of many of the world's cultures, interweaving widely diverse instruments and elements with extraordinary voices from what he refers to as "the greater symphony of the Earth," including wolves, whales, eagles and several dozen other species of "wilderness musicians." Winter grew up in Altoona, Pennsylvania, and was educated at Northwestern University, where he formed the Paul Winter Sextet in 1961. The sextet won the Intercollegiate Jazz Festival and was sent by the U.S. State Department on a six-month tour of 23 countries in Latin America. With his various ensembles over the years he has recorded 37 albums. His work has been honored with the Global 500 Award from the United Nations, the Award of Excellence from the United Nations Environment Program, the Joseph Wood Krutch Medal from the U.S. Humane Society, and four Grammy Awards.

Mary Youngblood, an American Indian musician, artist and poet, earned a 2003 Grammy Award for Best Native American Album for *Beneath the Raven Moon*. She was the 1999 and 2000 Native

American Music Awards Nammy winner for Flutist of the Year, as well as the 2000 Nammy's winner for Best Female Artist. Youngblood's solo debut album, titled *The Offering,* on Silver Wave Records, received rave reviews. In 2000, her second album, *Heart of the World,* won the Nammy Award for Best New Age Recording and placed at the top of Amazon.com's annual list of "The Best New Age CDs." At the 2002 Native American Music Awards Youngblood was nominated for Artist of the Year, Flutist of the Year and Best New Age Recording, for *Beneath the Raven Moon.* Her music is featured on more than 20 compilation albums, including the soundtrack for the popular IMAX film *Wolves* and the films *Naturally Native* and *Johan Padan.* Half Aleut and half Seminole, Youngblood says her songs come from "those who walked before me."

Eugenia Zukerman is a distinguished international flutist, recording artist, author and television correspondent for *CBS News Sunday Morning.* She was born in Cambridge, Massachusetts, and educated at Barnard College and The Juilliard School. Zukerman has been a popular performer for over 25 years, renowned for her elegant sound, lyrical phrasing, extraordinary agility and graceful stage presence. Zukerman continues to play with orchestras, in solo and duo recitals and in chamber music ensembles in North America, Europe and Asia. A versatile artist, she is also artistic director of the international Vail Valley Music Festival. Zukerman currently records exclusively for Delos. As an author, her fourth book, *In My Mother's Closet: An Invitation to Remember,* a collection of memories from people such as Claire Bloom, Renée Fleming, Carrie Fisher and Joy Behar, is scheduled for publication in April 2003 by Sorin Books.

Ellen Taaffe Zwilich is a classical composer whose works are enjoyed by diverse audiences the world over. Prolific in all musical forms save opera, Zwilich combines craft and inspiration to reflect an optimistic and humanistic spirit in her compositions. In 1983 her *Symphony No. 1* was honored with the Pulitzer Prize, bringing Zwilich, the first woman to win a Pulitzer in music, instantly into international focus. Since then her work has been commissioned and performed by nearly all the leading American and European ensembles and musicians. Each of Zwilich's first three symphonies has garnered her a Grammy nomination, as has her *Flute Concerto.* She has also been honored with a Guggenheim Fellowship, induction into the American Academy of Arts and Letters and *Musical America*'s "Composer of the Year" designation in 1999. In 1995, she was named to the first Composer's Chair in the history of Carnegie Hall.

Acknowledgments

The original pieces published in *Water Music* are held in copyright by the artists themselves. The copyright to each artist is as follows:

Samuel Adler's contribution, © Samuel Adler 2003

Vladimir Ashkenazy's contribution, © Vladimir Ashkenazy 2003

Emanuel Ax's contribution, © Emanuel Ax 2003

Keola Beamer's contribution, © Keola Beamer 2003

Jane Ira Bloom's words, © Jane Ira Bloom 2003

Dave Brubeck's contribution, © Dave Brubeck 2003

Sarah Chang's contribution, © Sarah Chang 2003

Bruce Cockburn's words, © Bruce Cockburn 2003

Basia Danilow's contribution, © Basia Danilow and Peter Sanders 2003

Renée Fleming's contribution, © Renée Fleming 2003

Pamela Frank's contribution, © Pamela Frank 2003

Alan Gampel's contribution, © Alan Gampel 2003

Evelyn Glennie's contribution, © Evelyn Glennie 2003

Mike Gordon's contribution, © Mike Gordon 2003

David Harrington's contribution, © David Harrington 2003

Mickey Hart's contribution, © 360° Productions 2003

Tilla Henkins' contribution, © Tilla Henkins 2003

Sharon Isbin's contribution, © Sharon Isbin 2003

Paavo Järvi's contribution, © Paavo Järvi 2003

Wesley Jefferson's contribution, © Wesley Jefferson 2003

Tunde Jegede's contribution, © Tunde Jegede 2003

Jaime Laredo's contribution, © Sharon Robinson and Jaime Laredo 2003

Iain Mac Harg's musical composition, © Iain Mac Harg 2002

Taj Mahal's contribution, © Taj Mahal 2003

Carol Maillard's contribution, © Carol Lynn Maillard 4 Jagadish Music 2003

Mischa Maisky's contribution, © Mischa Maisky 2003

Mary Elizabeth's contribution, © Mary Elizabeth/Voice of the Phoenix 2003

Bobby McFerrin's contribution, © Bobby McFerrin 2003

Brad Mehldau's contribution, © Brad Mehldau 2003

Susanne Mentzer's contribution, © Susanne Mentzer 2003

Midori's contribution, © Midori 2003

Yousuke Miura's contribution, © Yousuke Miura 2003

Keb' Mo's contribution, © Kevin R. Moore 2003

Randy Newman's contribution, © Randy Newman 2003

Nerissa Nield's contribution, © Nerissa Nields 2003

Erik Nielsen's words, © Erik Nielsen 2003; Erik Nielsen's musical composition, © Middle Branch Music 2003

Mark O'Connor's contribution, © Mark O'Connor 2003

Garrick Ohlsson's contribution, © Garrick Ohlsson 2003

Christopher Parkening's contribution, © Christopher Parkening 2003

Hermeto Pascoal's contribution, © Hermeto Pascoal 2003

Marcus Roberts' contribution, © Marcus Roberts 2003

Sharon Robinson's contribution, © Sharon Robinson and Jaime Laredo 2003

Paula Robison's contribution, © Paula Robison 2003

John Ruskey's contribution, © John Ruskey 2003

Nadja Salerno-Sonnenberg's contribution, © Nadja Salerno-Sonnenberg 2003

Peter Sanders' contribution, © Basia Danilow and Peter Sanders 2003

Joseph Schwantner's contribution, © Joseph Schwantner 2003

Russell Sherman's contribution, © Russell Sherman 2003

Eugene Skeef's contribution, © Eugene Skeef FRSA 1995

Richard Stoltzman's contribution, © Richard Stoltzman 2003

Tan Dun's contribution, © Tan Dun 2003

Gregory Turay's contribution, © Gregory Turay 2003

Randy Weston's contribution, © Randy Weston 2003

George Winston's contribution, © George Winston 2003

Paul Winter's contribution, © Paul Winter 2003

Mary Youngblood's contribution, © Mary Youngblood 2003

Eugenia Zukerman's contribution, © Eugenia Zukerman 2003

Ellen Taaffe Zwilich's contribution, © Ellen Taaffe Zwilich 2003

For permission to reprint previously published material, grateful acknowledgment is made to the following:

"Let It Rain"
by Patricia Barber
© 1997 Patricia Barber BMI
Used by Permission

"Pacific"
by Jane Ira Bloom
© Outline Music 1999
Used by Permission

Paul Winter   Carol Maillard   David Harrington

Nadja Salerno-Sonnenberg   Bruce Cockburn

Bobby McFerrin   Mary Elizabeth   Randy

Newman   Gordon   Beamer

Hermeto Pascoal   erissa Nields

Alan Gampel   Iain   ary Young-

blood   Tunde Jege   ming

Russell Sherman   Patricia

DATE DUE

DEMCO, INC. 38-2931

Barber   Vladimir Ashkenazy   Garrick Ohlsson

Basia Danilow   Peter Sanders   Eugenia Zuker-

man   Pete Seeger   Keb' Mo'   John Ruskey

George Winston   Emanuel Ax   Dave Brubeck